SpringerBriefs in Statistics

More information about this series at http://www.springer.com/series/8921

Janusz L. Wywiał

Sampling Designs Dependent on Sample Parameters of Auxiliary Variables

Second Edition

 Springer

Janusz L. Wywiał
Department of Statistics, Econometrics
and Mathematics
University of Economics
Katowice, Poland

ISSN 2191-544X   ISSN 2191-5458   (electronic)
SpringerBriefs in Statistics
ISBN 978-3-662-63412-7   ISBN 978-3-662-63413-4   (eBook)
https://doi.org/10.1007/978-3-662-63413-4

This Springer imprint is published by the registered company Springer-Verlag GmbH,
DE part of Springer Nature.
The registered company address is: Heidelberger Platz 3, 14197 Berlin, Germany

For Eva and Peter

Preface

We can observe gradual increases in the range of survey sampling methods and their applications in the fields of economics, geology, and agriculture, among others. These applications have been followed by new theoretical solutions that provide better sampling designs or estimators. Recently, several highly important properties of sampling designs have been discovered, and many new methods have been published. Computerization of human activity leads to a very rapid growth of complete databases on various populations. The resources can be treated as the values of the so-called auxiliary variables observed in the entire population. It is well known that the population sampling designs constructed on the basis of these data lead to a significant improvement in the accuracy of the estimation of the parameters of the variables under study. New developments in the construction of such plans are the subject of this book. Apart from the classical sampling designs used to estimate the parameters of discrete variables, continuous sampling plans useful for the estimation of the parameters of continuous populations are also considered. The proposed conditional sampling deigns provide new possibilities for accurate estimation of quantiles of the variable under study but also for means in domains. This is indicated by simulation studies of the accuracy of estimation of the parameters of the studied variables.

This small monograph is the result of the author's interest in survey designs. In general, the book represents a synthesis of contributions on sampling designs that are dependent on sample moments or the order statistics of auxiliary variables.

The book should be useful for students and statisticians whose work involves survey sampling. The book can offer new inspirations for those looking for new sampling designs dependent on auxiliary variables.

The Author is grateful to Reviewers for valuable comments.

This book is a result of the research supported by the grants: N N111 434137 from the Polish Ministry of Science and Higher Education, DEC-2012/07/B/H/03073 and 2016/21/B/HS4/00666 from National Scientific Center, Poland.

Katowice, Poland   Janusz L. Wywiał
March 2021

Contents

1   Introduction and Basic Sampling Strategies   1
  1.1   Aim and Outline of This Book   1
  1.2   Population and Parameters   2
  1.3   Sampling Design and Sampling Scheme   4
  1.4   Estimation   6
  References   12

**2   Sampling Designs Dependent on Sample Moments of Auxiliary
    Variables**   15
  2.1   Sampling Design Proportional to Sample Mean   15
  2.2   Sampford's Sampling Design   17
  2.3   Sampling Design Proportional to Sample Variance   18
  2.4   Sampling Designs Proportional to the Generalized Variance   20
  References   28

**3   Sampling Designs Based on Order Statistics of an Auxiliary
    Variable**   31
  3.1   Basic Properties of Order Statistics   31
  3.2   Sampling Design Proportional to Function of One-Order
        Statistic   33
  3.3   Sampling Design Proportional to Function of Two-Order
        Statistics   38
  3.4   Sampling Design Proportional to Function of Three-Order
        Statistics   46
  References   49

4   Simulation Analysis of the Efficiency of the Strategies   51
  4.1   Description of the Simulation Experiments   51
  4.2   Efficiency of Estimation Strategies Dependent on Sample
        Moments or Order Statistics   54
  4.3   Efficiency of Estimation Strategies Dependent on the Sum
        of Order Statistics   57
  4.4   Efficiency Estimation of Domain Mean   67

4.5   Estimation of Quantiles 70
4.6   Conclusions 77
References 78

**5   Sampling Designs Dependent on a Continuous Auxiliary
     Variable** 81
5.1   Introduction 81
5.2   Basic Definitions and Theorems 82
5.3   The Inclusion Function Proportional to the Values
      of the Auxiliary Variable 86
5.4   Sampling Designs as a Function of an Order Statistic 88
      5.4.1   General Results 88
      5.4.2   Uniform Distribution 92
5.5   Accuracy Analysis 100
References 107

Chapter 1
Introduction and Basic Sampling Strategies

Abstract This chapter provides an introduction to the problem of using auxiliary variables to support the estimation of the mean in finite and fixed populations. The basic definitions connected with sampling design are presented and properties of the well-known ratio, regression, and Horvitz–Thompson-type estimators are considered.

1.1 Aim and Outline of This Book

In this book, we consider estimates of the population parameters in finite and fixed populations, focusing on the population average of the variable under study. The values of the variable under study are observed in random samples that are selected according to pre-assigned sampling designs and sampling schemes. We assume that the values of auxiliary variables are observed in the whole population. The population average is estimated by means of a sampling strategy which is the pairing of an estimator and a sampling design. Extensive reviews of sampling designs or sampling schemes dependent on an auxiliary variable have been presented, for example, by Brewer and Hanif (1983), Chaudhuri and Vos (1988), Singh (2003), Chaudhuri and Stenger (2005), Tillé (2006), and in the International Encyclopedia (2011).

Our considerations are focused on particular sampling strategies that are dependent on sample moments or order statistics of a positively valued auxiliary variable. The main purpose of this brief monograph is to present the basic properties of such sampling strategies. Moreover, their accuracy will be analyzed.

We start with basic notations and the presentation of ordinary sampling strategies. In particular, the well-known Horvitz–Thompson (1952) statistic, as well as ratio or regression estimators, are considered. The class of strategies dependent on sample moments of auxiliary variables are presented in Chap. 2. The sampling designs proportional to the sample mean or sample variance are defined. The last design

© The Author(s), under exclusive license to Springer-Verlag GmbH, DE, part of Springer Nature 2021
J. L. Wywiał, *Sampling Designs Dependent on Sample Parameters of Auxiliary Variables*, SpringerBriefs in Statistics,
https://doi.org/10.1007/978-3-662-63413-4_1

is generalized to a sampling design proportional to the determinant of the sample variance–covariance matrix of the auxiliary variables. The new class of sampling designs that are proportional to functions of order statistics of the auxiliary variable are introduced in Chap. 3. In particular, the sampling design proportional to the value of the positive order statistic is considered. The inclusion probabilities of the sampling designs dependent on sample moments as well as order statistics are derived. Several new estimators dependent on sample moments or order statistics are also considered. A computer simulation-based comparison of accuracy of the defined sampling strategies is presented in Chap. 4.

In this second edition of the book, there are the additional considerations of the estimation parameters of a continuous random variable based on samples obtained via a continuous sampling design which was dependent on an auxiliary variable. The results are presented in Chap. 5.

In general, the conclusions allow us to choose the sampling strategy for particular situations, as determined by population distributions of a variable under study and auxiliary variables.

1.2   Population and Parameters

A finite population is denoted by U and it is a collection of N units where N is the size of the population and fixed. The members of the population U are usually denoted by the natural numbers $1, 2, \ldots, N$ and properties referring to the kth member of the population are indicated by a subscript k. The variable under study is denoted by y and its value, attached to the kth member of the population, is denoted by $y_k, k \in U$. The following parameters will be considered. The mean value is as follows:

$$\bar{y} = \frac{1}{N} \sum_{k \in U} y_k.$$

The total value of the variable y is

$$\tilde{y} = \sum_{k \in U} y_k = N\bar{y}.$$

If the variable y can take only the values 0 and 1, the number of values of the variable y equal to 1 is denoted by

$$M = \tilde{y}.$$

The ratio of the latter to the size of the population (fraction) is denoted by

$$w = \frac{M}{N}.$$

The variance is

$$v(y) = \frac{1}{N-1} \sum_{k=1}^{N} (y_k - \bar{y})^2.$$

The vector of auxiliary variables will be denoted by $x = [x_1...x_m]$. Let x_{kj} be the kth value of the jth variable, $k = 1, ..., N$, $j = 1, ..., m$. The matrix of variances and covariances of auxiliary variables will be denoted by $\mathbf{V}(x) = [v(x_t, x_j)]$, $t, j = 1, ..., m$, where

$$v(x_t, x_j) = \frac{1}{N-1} \sum_{k=1}^{N} (x_{kt} - \bar{x}_t)(x_{kj} - \bar{x}_j), \qquad \bar{x}_t = \frac{1}{N} \sum_{i=1}^{N} x_{it}.$$

The variances of the variable x_j are defined by the expression $v(x_j) = v(x_j, x_j)$. The covariance between the variable under study and an auxiliary variable is

$$v(x_j, y) = \frac{1}{N-1} \sum_{k=1}^{N} (x_{kj} - \bar{x}_j)(y_k - \bar{y}), \qquad j = 1, ..., m.$$

The correlation matrix of auxiliary variables will be denoted by $\mathbf{R} = \mathbf{R}(x) = [r(x_t, x_j)]$, $t, j = 1, \ldots, m$, where

$$r(x_t, x_j) = \frac{v(x_t, x_j)}{\sqrt{v(x_t)v(x_j)}}$$

In several cases, the mixed-central moments will be considered which are defined as follows:

$$v_{u,z}(x, y) = \frac{1}{N-1} \sum_{k=1}^{N} (x_k - \bar{x})^u (y_k - \bar{y})^z, \qquad u = 0, 1, 2, ... \quad z = 0, 1, 2, ...$$

Particularly,

$$v_{u,0}(x, y) = v_u(x), \qquad v_{2,0}(x, y) = v_2(x) = v(x), \qquad v_{1,1}(x, y) = v(x, y).$$

Let $\mathbf{X} = [x_{kj}]$ be the matrix of the dimensions $N \times m$ where x_{kj} is the kth observation of a jth auxiliary variable, $k = 1, ..., N$, $j = 1, ..., m$. Moreover, let $x_{k*} = [x_{k1}...x_{km}]$ be the row vector of the matrix $\mathbf{X}$. It is the kth observation of the values of the auxiliary variables (attached to the kth population element, $k = 1, ..., N$). From the geometrical point of view, the vector x_{k*} can be treated as the vector of coordinates of a point in the m-dimensional Euclidian space. The generalized variance of an m-dimensional variable is defined by Wilks (1932) as the determinant of the variance–covariance matrix:

$$g(y) = \det(\mathbf{V}(x)) \tag{1.1}$$

Let $q(x_{k_1*}, \ldots, x_{k_m*}, x_{k_{m+1}*})$ be the measure (volume) of the p-dimensional parallelotop spanned by the vectors that are all attached to the point $x_{k_{m+1}*}$, and the ends of these vectors are at the appropriate points $x_{k_1*}, \ldots, x_{k_m*}$. The volume of the parallelotop is determined by the following equation, see, e.g. Borsuk (1969):

$$q(x_{k_1*}, \ldots, x_{k_m*}, x_{k_{m+1}*}) = \left| \det \begin{bmatrix} x_{k_1*} - x_{k_{m+1}*} \\ \cdots\cdots\cdots\cdots \\ x_{k_m*} - x_{k_{m+1}*} \end{bmatrix} \right|$$

Let $q(x_{k_1*}, \ldots, x_{k_m*}, \bar{x})$ be the volume (measure) of the m-dimensional parallelotop spanned by the vectors that are all attached to the point $\bar{x}$ and their ends are at the points: $x_{k_1*}, \ldots, x_{k_m*}$.

A generalized variance $g(x)$ is proportional to the following sum of squared volumes spanned by the vectors that are all attached to the point $\bar{x}$ and whose ends have co-ordinates that are appropriate m-element combinations of rows of the matrix x (see, Anderson 1958, pp. 168–170):

$$g(x) = N^{-m} \sum_{\{k_1, \ldots, k_m\}} q^2(x_{k_1*}, \ldots, x_{k_m*}, \bar{x})$$

The generalized variance $g(x)$ is proportional to the following sum of squared volumes spanned by the $(m + 1)$ points whose co-ordinates are $(m + 1)$—element combinations of rows of the matrix $\mathbf{X}$, see e.g. Hardville (1997):

$$g(x) = N^{-m-1} \sum_{\{k_1, \ldots, i_{m+1}\}} q^2(x_{k_1*}, \ldots, x_{k_m*}, x_{k_{m+1}*})$$

The generalized variance is used as the coefficient measuring the scatter of observations of a multidimensional variable. When $g(x) = 0$, all observations of a m-dimensional variable are on not more than $(m - 1)$-dimensional hyperplane (see, e.g. Anderson 1958 or Hardville 1997).

1.3   Sampling Design and Sampling Scheme

The sample of the fixed size n, drawn from the population U will be denoted by s. Let $\mathbf{S}$ be the space sample. The function $P(s)$ on $\mathbf{S}$, satisfying

$$P(s) \geq 0 \text{ for all } s \in \mathbf{S} \text{ and } \sum_{s \in \mathbf{S}} P(s) = 1 \tag{1.2}$$

is called the sampling design.

The probability of selecting the fixed unit k to a sample s is called the inclusion probability of the first order and denoted by π_k. It is determined by the following expression:

$$\pi_k = \sum_{s:k \in s} P(s), \quad k = 1, \ldots, N.$$

Similarly, the second-order inclusion probability is as follows:

$$\pi_{kl} = \sum_{k,l \in s} P(s), \quad k \neq l, \quad k = 1, \ldots, N, \quad l = 1, \ldots, N.$$

The sampling design of the simple random sample drawn without replacement is as follows:

$$P_0(s) = \binom{N}{n}^{-1} \quad \text{for all} \quad s \in \mathbf{S} \tag{1.3}$$

The inclusion probabilities for the sampling design P_0 are as follows:

$$\pi_k = \frac{n}{N} \quad \pi_{kl} = \frac{n(n-1)}{N(N-1)} \tag{1.4}$$

The set of probabilities implementing a sampling design P_0 is defined as described hereafter: The probability of selecting a fixed population element k_1 in the sample is as follows:

$$p(k_1) = N^{-1} \text{ for } k_l = 1, \ldots, N$$

The conditional probability of selecting the fixed population element k_i in a sample, provided that the elements $k_{i-1}, \ldots, k_1$ have just been selected in the sample, is as follows:

$$p(k_i | k_{i-1}, \ldots, k_1) = \frac{1}{N-i+1} \text{ for } i = 2, \ldots, n \text{ and } k_i = 1, \ldots, N$$

Let us note that Rao (1962) proved that for any given design $P(s)$ there exists at least one sampling scheme that implements $P(s)$.

1.4   Estimation

Let t_s be an estimator of a population parameter denoted by $\theta \in \Theta$ where Θ is the parameter's space. The estimation of the parameter will be considered under the various estimators and sampling designs $P(s)$. That is why according to the definition of Cassel Särndal and Wretman (1977), p. 26, it is convenient to treat the pair $(t_s, P(s))$ as the sampling (estimation) strategy of θ. The strategy $(t_s, P(s))$ is unbiased if

$$E(t_s, P(s)) = \theta \quad \text{for all} \quad \theta \in \Theta$$

where

$$E(t_s, P(s)) = \sum\nolimits_{s \in S} t_s P(s)$$

The variance of the strategy, per Cassel Särndal and Wretman (1977), is as follows:

$$V(t_s, P(s)) = \sum_{s \in S} (t_s - E(t_s))^2 P(s)$$

The mean square error of the strategy is denoted by

$$MSE(t_s, P(s)) = E((t_s - \theta)^2, P(s)) = V(t_s, P(s)) + b^2$$

where $b = E(t_s, P(s)) - \theta$ is the bias and

$$E((t_s - \theta)^2, P(s)) = \sum_{s \in S} (t_s - \theta)^2 P(s).$$

Let us underline that we are mainly concerned with estimation of the mean value of a variable under study.

Simple Random Sample Mean

The well-known simple random sample mean is the unbiased strategy for a population mean. So, $E(\bar{y}_s, P_0(s)) = \bar{y}$ where

$$\bar{y}_s = \frac{1}{N} \sum_{i \in s} y_i \tag{1.5}$$

The variance of the estimation strategy $(\bar{y}_s, P_0(s))$ is

$$V(\bar{y}_s, P_0) = \frac{N - n}{Nn} v(y) \tag{1.6}$$

An unbiased estimator of the variance $v(y)$ is

$$V_s(\bar{y}_s, P_0) = \frac{N-n}{Nn} v_s(y) \tag{1.7}$$

$$v_s(y) = \frac{1}{n-1} \sum_{k \in s} (y_k - \bar{y}_s)^2$$

Let $(t_s, P(s))$ be a sampling strategy for a population mean. The relative efficiency coefficient for the estimation of the population mean, per Kish (1965), is defined as follows:

$$deff(t_s, P(s)) = \frac{MSE(t_s, P(s))}{V(\bar{y}_s, P_0(s))} \tag{1.8}$$

Ratio Estimators

Let us consider the following ratio estimator from simple random sample.

$$\bar{y}_{rs} = \frac{\bar{y}_s}{\bar{x}_s} \bar{x} \tag{1.9}$$

Its variance is

$$V(\bar{y}_{rs}) \approx \frac{N-n}{nN(N-1)} \sum_{k=1}^{N} (y_k - hx_k)^2 \tag{1.10}$$

where $h = \frac{\bar{y}}{\bar{x}}$ or

$$V(\bar{y}_{rs}) \approx \frac{N-n}{nN} v(y) \left[1 + \frac{\gamma^2(x)}{\gamma^2(y)} - 2\frac{\gamma(x)}{\gamma(y)} r(x, y) \right] \tag{1.11}$$

where $\gamma(x) = \frac{\sqrt{v(x)}}{\bar{x}}$, $\gamma(y) = \frac{\sqrt{v(y)}}{\bar{y}}$.

The bias of the estimator $\bar{y}_{rs}$ is as follows (see e.g. Cochran 1963, p. 161 or Konijn 1973, p. 100):

$$b_r = E(\bar{y}_{rs}) - \bar{y} \approx \frac{N-n}{Nn} \sqrt{v(y)} \gamma(x) \left[\frac{\gamma(x)}{\gamma(y)} - r(x, y) \right]$$

Thus, the statistic $\bar{y}_{rs}$ is an asymptotically unbiased estimator for $\bar{y}$. If the correlation coefficient $r(x, y) > 0$, the relative bias of the estimator is (see e.g. Cochran 1963, p. 162 and Konijn 1973, p. 102):

$$\frac{|b_r|}{\sqrt{V(\bar{y}_{rs})}} \le \gamma(x)$$

Therefore, a non-negative auxiliary variable should be chosen in such a way that its variation coefficient has to be as small as possible and the correlation coefficient $r(x, y)$ should be as high as possible.

The variance of the statistic $\bar{y}_{rs}$ can be estimated by means of the following two statistics:

$$V_s(\bar{y}_{rs}) = \frac{N-n}{nN(n-1)} \sum_{k \in s}(y_k - h_s x_k)^2 \tag{1.12}$$

where

$$h_s = \frac{\bar{y}_s}{\bar{x}_s}$$

or

$$\bar{V}_s(\bar{y}_{rs}) = \frac{N-n}{nN}[v_s(y) + h_s^2 v_s(x) - 2h_s v_s(x, y)]$$

where

$$v_s(x, y) = \frac{1}{n-1}\sum_{k \in s}(x_k - \bar{x}_s)(y_k - \bar{y}_s), \qquad v_s(y) = v_s(y, y).$$

The relative efficiency coefficient is as follows:

$$e_w(\bar{y}_{rs}/\bar{y}_s) = \frac{V(\bar{y}_{rs})}{V(\bar{y}_s)} = 1 + \frac{\gamma(x)}{\gamma(y)}\left[\frac{\gamma(x)}{\gamma(y)} - 2r(x, y)\right] \tag{1.13}$$

Therefore, the estimator $\bar{y}_{rs}$ is more accurate than $\bar{y}_s$, when

$$r(x, y) > \frac{1}{2}\frac{\gamma(x)}{\gamma(y)} \text{ and } \gamma(x) < 2\gamma(y).$$

Regression Estimator

The well-known ordinary regression estimator from the simple random sample is as follows:

$$\bar{y}_{regs} = \bar{y}_s + a_s(\bar{x}_s - \bar{x}) \tag{1.14}$$

where a_s is the slope coefficient:

$$a_s = \frac{v_s(x, y)}{v_s(x)} \tag{1.15}$$

The statistic $\bar{y}_{regs}$ is an almost unbiased estimator of the mean $\bar{y}$. Its bias is approximately as follows:

$$b_{reg} \approx \frac{1}{n}\frac{N-n}{N-2}\sqrt{v(x)(\beta_2-1)}(\theta_{21}(x,y) - \theta_3(x)r(x,y))$$

and

$$|b_{reg}| \leq \frac{2}{n}\sqrt{v(x)(\beta_2(x)-1)}$$

where

$$\beta_2(x) = \frac{c_4(x)}{v^2(x)}, \qquad \theta_{21}(x,y) = \frac{v_{21}(x,y)}{\sqrt{v(y)(c_4(x)-v^2(x))}},$$

$$\theta_3(x) = \frac{v_3(x)}{\sqrt{v(x)(c_4(x)-v^2(x))}}, \qquad |\theta_3(x)| \leq 1, \quad |\theta_{21}(x,y)| \leq 1.$$

The well-known kurtosis coefficient is denoted by $\beta_2(x)$. The parameter $\theta_3(x)$ has been derived as the correlation coefficient of the sample mean and the sample variance, see Kendall and Stuart (1967) for a large sample size. Moreover, let us note that the parameter $\theta_3(x)$ can be defined as the correlation coefficient between the variables x and $(x-\bar{x})^2$, whereas the parameter $\theta_{21}(x,y)$ was defined as the correlation coefficient between the variables y and $(x-\bar{x})^2$. Hence, $\theta_3(x)$ is also treated as the skewness coefficient of a variable x, too.

The variance of the estimator is approximately given by the expression:

$$V(\bar{y}_{regs}) \approx \frac{N-n}{Nn}v(y)(1-r^2(x,y)). \tag{1.16}$$

The estimator of the variance is

$$V_s(\bar{y}_{regs}) = \frac{N-n}{Nn}v_s(y)(1-r_s^2(x,y)). \tag{1.17}$$

where

$$r_s(x,y) = \frac{v_s(x,y)}{\sqrt{v_s(x)v_s(y)}}.$$

The relative efficiency coefficient is

$$e_w(\bar{y}_{regs}/\bar{y}_s) \approx V(\bar{y}_{regs}/V\bar{y}_s) = 1 - r^2(x,y). \tag{1.18}$$

Hence, the regression estimator is not less precise than the simple random sample mean.

Moreover, let us note that in the survey sampling literature some modifications of the ratio, regression, and so-called product estimators are considered (see, e.g. Kadilar and Cingi 2006, Kumar and Kadilar 2013, Murthy 1964, Singh et al. 2011, Srivenkataramana 1980, Swain 2013). Those estimators are not taken into account in the next chapters because they are determined on the basis of the simple random sample.

In the case when a multivariate auxiliary variable is observed in the simple random sample drawn without replacement, the ratio estimators was generalized by Olkin (1958) and the regression estimator was considered by Hung (1985) and Wywiał (2003). In the next chapter, the multivariate regression estimator is taken into account under the sampling design proportional to the determinant of the sample variance–covariance matrix of auxiliary variables.

Horvitz–Thompson Estimator

Let us assume that a sample s of the size n is drawn without replacement from a finite population. Let $I_k = 1$ ($I_k = 0$), if $k \in s$ (if $k \notin s$) for all $k = 1, \ldots, N$. It is well known that

$$\pi_k = E(I_k), \quad \pi_{ki} = E(I_k I_i), \quad \sum_{k=1}^{N} \pi_k = n,$$
$$V(I_k) = \pi_k(1 - \pi_k), \quad V(I_k, I_i) = \pi_{ki} - \pi_k \pi_i \, for \, k \neq i.$$

Horvitz and Thompson (1952) proposed the following estimator of the population average $\bar{y}$:

$$\bar{y}_{HTs} = \frac{1}{N} \sum_{k=1}^{N} \frac{I_k y_k}{\pi_k} = \frac{1}{N} \sum_{k \in s} \frac{y_k}{\pi_k} \tag{1.19}$$

The statistic $\bar{y}_{HTs}$ is the unbiased estimator of the population mean $\bar{y}$ when $\pi_k > 0$ for all $k = 1, \ldots, N$. Its variance is determined by the following expression:

$$V(\bar{y}_{HTs}) = \frac{1}{N^2} \sum_{k=1}^{N} \left(\frac{y_k}{\pi_k} \right)^2 \pi_k(1 - \pi_k) + \frac{2}{N^2} \sum_{i=1}^{N} \sum_{k>i}^{N} \frac{y_k y_i}{\pi_k \pi_i}(\pi_{ki} - \pi_i \pi_k) \tag{1.20}$$

For a fixed sample size,

$$V(\bar{y}_{HTs}) = \frac{1}{N^2} \sum_{i=1}^{N} \sum_{k>i}^{N} (\pi_k \pi_i - \pi_{ki}) \left(\frac{y_k}{\pi_k} - \frac{y_i}{\pi_i} \right)^2 \tag{1.21}$$

The variance $V(\bar{y}_{HTs})$ is estimated by means of the following statistic:

$$V_s(\bar{y}_{HTs}) = \frac{1}{N^2} \sum_{k=1}^{N} I_k \left(\frac{y_k}{\pi_k}\right)^2 (1 - \pi_k) + \frac{2}{N^2} \sum_{i=1}^{N} \sum_{k>i}^{N} I_k I_i \frac{y_k y_i}{\pi_k \pi_i} \frac{\pi_{ki} - \pi_k \pi_i}{\pi_{ki}}$$

$$(1.22)$$

This statistic is the unbiased estimator of the variance $V(\bar{y}_{HTs})$ but it can take negative values.

When the sample size is fixed and $\pi_k \pi_i - \pi_{ki} > 0$ for each $k \neq i = 1, \ldots, N$ the non-negatively valued estimator is as follows, (see Sen 1953, Yates and Grundy 1953):

$$\bar{V}_s(\bar{y}_{HTs}) = \frac{1}{N^2} \sum_{i=1}^{N} \sum_{k>i}^{N} I_k I_i \frac{\pi_k \pi_i - \pi_{ki}}{\pi_{ki}} \left(\frac{y_k}{\pi_k} - \frac{y_i}{\pi_i}\right)^2 \qquad (1.23)$$

Moreover, let us note that Horvitz and Thompson proposed the estimator of the population average in the case when a sample is selected with replacement.

The Horvitz–Thompson estimator can be the component of the ratio and regression type estimators, (see e.g. Särndal et al. 1992). The ratio estimator is as follows:

$$\bar{y}_{rHTs} = \bar{y}_{HTs} \frac{\bar{x}}{\bar{x}_{HTs}}, \qquad (1.24)$$

It is an approximately unbiased estimator of the population mean and its variance is approximately as follows:

$$V(\bar{y}_{rHTs}) \approx V(\bar{y}_{HTs}) + h^2 V(\bar{x}_{HTs}) - 2h Cov(\bar{x}_{HTs}, \bar{y}_{HTs}) \qquad (1.25)$$

where

$$h = \frac{\bar{y}}{\bar{x}}, \qquad V(\bar{x}_{HTs}) = Cov(\bar{x}_{HTs}, \bar{x}_{HTs}),$$

$$Cov(\bar{x}_{HTs}, \bar{y}_{HTs}) = \frac{1}{N^2} \sum_{k=1}^{N} \frac{x_k y_k (1 - \pi_k)}{\pi_k} + \frac{2}{N^2} \sum_{i=1}^{N} \sum_{k>i}^{N} \frac{x_k y_i}{\pi_k \pi_i} (\pi_{ki} - \pi_i \pi_k).$$

$$(1.26)$$

The variance $V(\bar{y}_{rHTs})$ is estimated by means of the following statistic:

$$V_s(\bar{y}_{rHTs}) = V_s(\bar{y}_{HTs}) + h_s^2 V_s(\bar{x}_{HTs}) - 2h_s Cov_s(\bar{x}_{HTs}, \bar{y}_{HTs}) \qquad (1.27)$$

where $h_s = \frac{\bar{y}_{HTs}}{\bar{x}_{HTs}}$,

$$Cov_s(\bar{x}_{HTs}, \bar{y}_{HTs}) = \frac{1}{N^2} \sum_{k=1}^{N} \frac{x_k y_k (1 - \pi_k)}{\pi_k^2} I_k + \frac{1}{N^2} \sum_{i=1}^{N} \sum_{k \neq i}^{N} \frac{x_k y_i}{\pi_k \pi_i} \frac{\pi_{ki} - \pi_i \pi_k}{\pi_{ki}} I_k I_i.$$

The regression type estimator is as follows:

$$\bar{y}_{regHTs} = \bar{y}_{HTs} + a_{HTs}(\bar{x} - \bar{x}_{HTs}) \qquad (1.28)$$

where

$$a_{HTs} = \frac{v_{HTs}(x, y)}{v_{HTs}(x)} \tag{1.29}$$

$$v_{HTs}(x, y) = \frac{1}{N-1} \sum_{k=1}^{N} (x_k - \bar{x}_{HTs})(y_k - \bar{y}_{HTs}) \frac{I_k}{\pi_k}, \qquad v_{HTs}(x) = v_{HTs}(x, x). \tag{1.30}$$

The statistic $\bar{y}_{reg HTs}$ is the approximately unbiased estimator of the population mean and its variance approximately shows the following expressions:

$$V(\bar{y}_{reg HTs}) \approx V(\bar{y}_{HTs}) + a^2 V(\bar{x}_{HTs}) - 2a Cov(\bar{x}_{HTs}, \bar{y}_{HTs}) \tag{1.31}$$

where $a = \frac{c_*(x,y)}{v_*(x)}$.
The unbiased estimator of the variance $V(\bar{y}_{reg HTs})$ is as follows:

$$V_s(\bar{y}_{reg HTs}) = V_s(\bar{y}_{HTs}) + a_{HTs}^2 V_s(\bar{x}_{HTs}) - 2a_{HTs}^2 Cov_s(\bar{x}_{HTs}, \bar{y}_{HTs}) \tag{1.32}$$

Let us note that more properties of the Horvitz–Thompson estimator are considered e.g. by Barbiero and Mecatti (2010), Berger (1998), Hulliger (1995), Patel and Patel (2010), Rao et al. (1962), Rao (2004).

References

Anderson, T. W. (1958). *An Introduction to Multivariate Statistical Analysis*. New York: Wiley.

Barbiero, A., Mecatti, F. (2010). Bootstrap algorithms for variance estimation in PS sampling. In P. Mantovan, & P. Secchi (Eds.), *Complex Data Modeling and Computationally Intensive Statistical Methods* (pp. 57–70). Springer, Italia.

Berger, Y. G. (1998). Rate of convergence for asymptotic variance of the Horvitz Thompson estimator. *Journal of Statistical Planning and Inference, 74*, 149–168.

Borsuk, K. (1969). *Multidimensional Analytic Geometry*. Warsaw: PWN.

Brewer, K. R. W., & Hanif, M. (1983). *Sampling with Unequal Probabilities*. New York-Heidelberg-Berlin: Springer.

Cassel, C. M., Särndal, C. E., & Wretman, J. W. (1977). *Foundation of Inference in Survey Sampling*. New York, London, Sydney, Toronto: Wiley.

Chaudhuri, A., & Stenger, H. (2005). *Survey Sampling. Theory and Methods* (2nd ed.). Boca Raton-London-New York-Singapore: Chapman & Hall CRC.

Chaudhuri, A., & Vos, J. W. E. (1988). *Unified Theory of Survey Sampling*. Amster-dam-New York-Oxford-Tokyo: North Holland.

Cochran, W. G. (1963). *Sampling techniques*. New York: Wiley.

Hardville, D. A. (1997). *Matrix Algebra From a Statistician's Perspective*. New York-Berlin-Heiderberg-Barcelona-Hong Kong-London-Milan-Paris-Singapore-Tokyo: Springer.

Horvitz, D. G., & Thompson, D. J. (1952). A generalization of the sampling without replacement from finite universe. *Journal of the American Statistical Association, 47*, 663–685.

Hulliger, B. (1995). Outlier robust Horvitz–Thompson estimator. *Survey Methodology, 21*, 79–87.

Hung, H. M. (1985). Regression estimators with transformed auxiliary variates. *Statistics & Probability Letters, 3,* 239–243.

Lovric, M. (2011). International Encyclopedia of Statistical Science. Springer, Berlin.

Kadilar, C., & Cingi, H. (2006). Improvement in estimating the population mean in simple random sampling. *Applied Mathematics Letters, 19,* 75–79.

Kendall, M. G., & Stuart, A. (1967). *The Advanced Theory of Statistics: Inference and Relationship* (Vol. 2). London: Charles Griffin and Company Limited.

Kish, L. (1965). *Survey Sampling.* New York- London-Sydney: Wiley.

Konijn, H. S. (1973). *Statistical Theory of Sample Survey and Analysis.* Amsterdam-London, New York: American Elsevier Publishing Company Inc, North-Holland Publishing Company Inc.

Kumar, S. Y., & Kadilar, C. (2013). Improved class of ratio and product estimators. *Applied Mathematics and Computation, 219,* 10726–10731.

Murthy, M. N. (1964). Product method of estimation. *Sankhya, Series A, 26,* 69–74.

Olkin I. (1958). Multivariate ratio estimation for finite populations. *Biometrika, 45,* 154–165.

Patel, P. A., & Patel, J. S. (2010). A Monte Carlo comparison of some variance estimators of the Horvitz–Thompson estimator. *Journal of Statistical Computation and Simulation, 80*(5), 489–502.

Rao, J. N. K., Hartley, H. O., & Cochran, W. G. (1962). On a simple procedure of unequal probability sampling without replacement. *Journal of the Royal Statistical Association B, 24*(2), 482–491.

Rao, T. J. (2004). Five decades of the Horvitz–Thompson estimator and furthermore. *Journal of the Indian Society of Agricultural Statistics, 58,* 177–189.

Rao, T.V.H. (1962). An existence theorem in sampling theory. *Sankhya A, 24,* 327–330.

Särndal, C. E., Swensson, B., & Wretman, J. (1992). *Model Assisted Survey Sampling.* New York-Berlin-Heidelberg-London-Paris-Tokyo-Hong Kong- Barcelona-Budapest: Springer.

Sen, A. R. (1953). On the estimate of variance in sampling with varying probabilities. *Journal of the Indian Society of Agicultural Statistics, 5*(2), 119–127.

Singh, S. (2003). *Advanced Sampling Theory with Applications. Volume I and Volume II.* Dordrecht, Boston, London: Kluwer Academic Publishers.

Singh, R., Kumar, M., Chauhan, P., Sawan, N., Smarandache, F.(2011). A general family of dual to ratio-cum-product estimator in surveys. *Statistics in Transition-New Series, 12*(3), 587–594.

Srivenkataramana, T. (1980). A dual to ratio estimator in sample surveys. *Biometrika, 67,* 199–204.

Swain, A. K. P. C. (2013). On some modified ratio and product type estimators-revisited. *Revisita Evista Investigacion Operacional, 34*(1), 35–57.

Tillé, Y. (2006). *Sampling Algorithms.* Berlin: Springer.

Wilks, S. S. (1932). Certain generalization in the analysis of variance. *Biometrika, 24,* 471–494.

Wywiał, J. L. (2003). *Some Contributions to Multivariate Methods in Survey Sampling.* Katowice University of Economics, Katowice. https://www.ue.katowice.pl/fileadmin/user_upload/wydawnictwo/Darmowe_E-Booki/Wywial_Some_Contributions_To_Multivariate_Methods_In_Survey_Sampling.pdf

Yates, F., & Grundy, P. M. (1953). Selection without replacement from within strata with probability proportional to size. *Journal of the Royal Statistical Society, Series B, 15,* 235–261.

Chapter 2
Sampling Designs Dependent on Sample Moments of Auxiliary Variables

This chapter describes the properties of ratio-type estimators from a sample drawn according to sampling design proportional to a sample mean of an auxiliary variable. The next sampling design proportional to the sample variance of an auxiliary variable is considered as a particular case of a sampling design dependent on the sample generalized variance of a multivariate auxiliary variable. Properties of regression-type estimators under this sampling design are also considered. The well-known Sampford sampling design is taken into account as a sampling design with inclusion probabilities proportional to the auxiliary variable values.

2.1 Sampling Design Proportional to Sample Mean

In the case of a simple random sample drawn without replacement, the ratio estimator can be better than the sample mean, as was shown in the previous chapter. However, the ratio estimator is biased, especially when the sample size is small. To eliminate the bias, Lahiri (1951), Midzuno (1952), and Sen (1953) proposed,[1] the following sampling design proportional to the sample mean.

Let $x = [x_1 \ldots x_N]$ be the vector of an auxiliary variable's observations and $x_k > 0$ for $k = 1, ..., N$. Let us recall that the sample and population means of the auxiliary variable are denoted by $\bar{x}_s = \frac{1}{n} \sum_{k \in s} x_k$ and $\bar{x} = \frac{1}{N} \sum_{k=1}^{N} x_k$, respectively. The size n of a sample is the effective sample size. The sampling design proportional to the sample mean of the auxiliary variable is as follows:

$$P_{LMS}(s) = \frac{1}{\binom{N}{n}} \frac{\bar{x}_s}{\bar{x}}. \tag{2.1}$$

---

[1] According to Chaudhuri and Stenger (2005), p. 25.

© The Author(s), under exclusive license to Springer-Verlag GmbH, DE, part of Springer Nature 2021
J. L. Wywiał, *Sampling Designs Dependent on Sample Parameters of Auxiliary Variables*, SpringerBriefs in Statistics, https://doi.org/10.1007/978-3-662-63413-4_2

The first- and second-order inclusion probabilities are as follows (see, e.g. Rao 1977):

$$\pi_k = \frac{N-n}{(N-1)N}\frac{x_k - \overline{x}}{\overline{x}} + \frac{n}{N} > \frac{n-1}{N-1}, \tag{2.2}$$

$$\pi_{jk} = \frac{n(n-1)}{N(N-1)} + \frac{(n-1)(N-n)}{(N-2)(N-1)N}\frac{x_k + x_j - 2\overline{x}}{\overline{x}} > \frac{n(n-1)}{N(N-1)},$$

where $k \neq j = 1, \ldots, N$.

The well-known enumerate sampling scheme (see, e.g. Tillé 2006) can be applied when implementing the sampling design, but it is not useful for a large population or sample size, which is why the following sampling scheme is applied. The first element is drawn from the population to the sample with probability $p_k = \frac{x_k}{\tilde{x}}$, $k = 1, \ldots, N$ where $\tilde{x} = \sum_{i \in U} x_i$. The subsequent elements of the sample are drawn without replacement from the remaining $N-1$ elements of the population as the simple random sample of size $n-1$.

Under the sampling design $P_{LMS}(s)$, the ordinary ratio estimator becomes an unbiased estimator of the population mean. Thus, $E(\overline{y}_{rs}, P_{LMS}(s)) = \overline{y}$ where $\overline{y}_{rs}$ is given by the expression (1.9).

Wywiał (2003) derived the following expression:

$$V(\overline{y}_{HTs}, P_{LMS}(s)) \approx V(\overline{y}_s, P_0) + \frac{v_2(y)}{n^2}(2\kappa\rho - \gamma_x \eta_{12} + \gamma_x \kappa \eta_{21} + \kappa^2 - \gamma_x^2 \rho^2) \tag{2.3}$$

where $V(\overline{y}_s, P_0)$ is explained by the expressions (1.3) and (1.5), $\kappa = \frac{\gamma_x}{\gamma_y}$, and

$$\eta_{rs} = \eta_{rs}(x, y) = \frac{v_{rs}}{v_2^{r/2}(x)v_2^{s/2}(y)}, \quad \eta_{11}(x, y) = \rho$$

Hence, when $N \to \infty$, $n \to \infty$ and $N - n \to \infty$,

$$V(\overline{y}_{HTs}, P_{LMS}(s)) \to V(\overline{y}_s, P_0).$$

Let us assume that (x_i, y_i), $i = 1, \ldots, N$ are outcomes of a two-dimensional normal random variable. In this case, the expression (2.3) is reduced to the following:

$$V(\overline{y}_{HTs}, P_{LMS}(s)) \approx \frac{N-n}{Nn}v(y) + \frac{v_2(y)}{n^2}(2\kappa\rho + \kappa^2 - \gamma_x^2 \rho^2).$$

Hence, particularly if $\kappa = 1$ and $\rho > \frac{2}{\gamma_x^2}$, the strategy $(\overline{y}_{HTs}, P_{LMS}(s))$ is more accurate than the simple random sample mean. More details about the properties of the strategies under the sampling design $P_{LMS}(s)$ are considered, e.g. by Rao (1966), Chaudhuri and Arnab (1981), and Srivenkataramana (2002).

Walsh (1970) proposed the following modification of the ratio estimator:

$$\bar{y}_{ws} = \frac{\bar{y}_s \bar{x}}{\bar{x} + A(\bar{x}_s - \bar{x})}.$$

Bhushan et al. (2009) generalized the sampling scheme of Lahiri–Midzuno–Sen in the following way: First, the kth population element is selected according to the probability:

$$p(k) = \frac{\bar{x} + A(x_k - \bar{x})}{N\bar{x}}, \qquad k = 1, .., N.$$

The subsequent elements of the sample are drawn without replacement from the remaining $N - 1$ elements of the population as the simple random sample of size $n - 1$. The estimator $\bar{y}_{ws}$ is unbiased under this sampling scheme. The variance of the estimator $\bar{y}_{ws}$ takes the minimal value equal to the variance of the regression estimator, given by the expression (1.16) under the generalized Lahiri–Midzuno–Sen sampling scheme when $A = \frac{\rho}{\kappa}$.

2.2 Sampford's Sampling Design

The well-known large class of sampling designs leads to selecting population units with inclusion probabilities proportional to the positive values of the auxiliary variable, so $\pi_k \propto x_k$, $k = 1, ..., N$. One of them was proposed by Sampford (1967)

$$P_{Sd}(s) = cn(1 - \bar{\pi}_s) \prod_{k \in s} \frac{\pi_k}{1 - \pi_k} \tag{2.4}$$

where

$$\bar{\pi}_s = \frac{1}{n} \sum_{k \in s} \pi_k, \qquad \pi_k = \frac{n x_k}{\sum_{i=1}^{N} x_i}, \qquad k = 1, ..., N,$$

The parameter c is determined in such a way that $\sum_{s \in S} P_{Sd}(s) = 1$.

The rejective sampling scheme is as follows: the first population unit is selected to the sample with probability π_k/n, $k = 1, ..., N$, and all the subsequent population units are drawn with probabilities proportional to $\frac{\pi_k}{1 - \pi_k}$, $i = 1, ..., N$, $i \neq k$. When any unit is selected multiple times to the sample s, the sample is rejected, and the sampling procedure is repeated. The exact expression for the parameter c, the inclusion probabilities of the second rank and many other properties of the sampling design can be found in the monograph by Tillé (2006).

Finally, let us note that a review of the considered methods of sampling with inclusion probabilities proportional to values of the auxiliary variable can be found, e.g. in the books by Tillé (2006) or Brewer and Hanif (1983).

2.3   Sampling Design Proportional to Sample Variance

In the case of the simple random sample drawn without replacement, the regression estimator is biased. To eliminate the bias, Singh and Srivastava (1980) proposed the following sampling design proportionate to the sample variance:

$$P_{SS}(s) = \frac{1}{\binom{N}{n}} \frac{v_s(x)}{v(x)} \tag{2.5}$$

where $v_s(x) = \frac{1}{n-1} \sum_{k \in s} (x_k - \bar{x}_s)^2$ and $v(x) = \frac{1}{N-1} \sum_{k=1}^{N} (x_k - \bar{x})^2$. Let

$$q_k = \frac{x_k - \bar{x}}{\sqrt{v_\#(x)}}, \qquad v_\#(x) = \frac{N-1}{N} v(x), \qquad k = 1, \ldots, N$$

The probabilities of inclusion are as follows:

$$\pi_k = \frac{n}{N} + \frac{N-n}{N(N-2)}(q_k^2 - 1) \geq \frac{(n-1)(N-1)}{N(N-2)} \tag{2.6}$$

$$\pi_{jk} = \frac{n(n-1)}{N(N-1)} + \frac{(N-n)}{N(N-2)}\left(1 - \frac{N-1)(N-n-1)}{N(N-3)}\right)(q_j^2 + q_k^2)+$$
$$- \frac{2(N-n)(N-n-1)}{N^2(N-2)(N-3)} q_j q_k - \frac{2(N-n)(N-n-1)}{N^2(N-1)(N-2)(N-3)}$$

for $j \neq k = 1, \ldots, N$.

Singh and Srivastava (1980) proposed a sampling scheme implementing the sampling design $P_{SS}(s)$. First, the following sequence of squared differences has to be determined:

$$\alpha_{ij} = (x_i - x_j)^2, \ j > i = 1, \ldots, N.$$

The first two population elements are selected to the sample proportionally to the values α_{ij}. The subsequent elements are selected as the simple random sample of size $n - 2$ drawn without replacement from the remaining population.

Based on the above sampling scheme, the following can be derived. The first element is selected for the sample with the probability:

$$p_k = \frac{1}{2N}\left(1 + \frac{(x_k - \bar{x})^2}{v_\#(x)}\right), \qquad k = 1, \ldots, N.$$

When the kth population element is selected, the second element is drawn with the following probability:

$$p_{j|k} = \frac{1}{N} \frac{(x_j - x_k)^2}{v_\#(x) + (x_k - \bar{x})^2}$$

where $j = 1, ..., N$, $k = 1, ..., N$, and $j \neq k$. Similarly, such as in the case of the previous sampling scheme, the next elements are selected as the simple random sample of size $n - 2$ drawn without replacement from the remaining population.

Singh and Srivastava (1980) proved that the ordinary regression estimator, as determined by the expression (1.14), becomes unbiased for the population mean under the sampling design $P_{SS}(s)$. Its variance is approximately equal to the right side of the equation (1.16).

The approximate variance of the strategy $(\bar{y}_{HTS}, P_{SS}(s))$ is as follows:

$$V(\bar{y}_{HTS}, P_{SS}(s)) \approx V(\bar{y}_S, P_0(s)) + \frac{\bar{y}^2}{n^2}(-\gamma_y^2(2\rho^2 + \eta_{22} + 1) + \gamma_y(\eta_{41} - 4\eta_{21}) + \eta_{40} - 1)$$

where t_{HTS} is the Horvitz–Thompson (1952) defined by the expression (1.19).

Under the assumption that $\eta_{22} = 1 + 2\rho^2$, $\eta_{12} = 0$ and $\eta_{14} = 0$ (e.g. in the case when variables (y, x) have an approximately two-dimensional normal distribution), the variance takes the following form:

$$V(\bar{y}_{HTS}, P_{SS}(s)) \approx V(\bar{y}_S, P_0(s)) + \frac{2\bar{y}^2}{n^2}(1 - \gamma_y^2(1 + 2\rho^2))$$

For sufficiently large n and N, the strategy $(t_{HTS}, P_{SS}(s))$ is better than $(\bar{y}_S, P_0(s))$ if $|\gamma_y| > 1$ or if $1 > |\gamma_y| > \frac{1}{\sqrt{3}}$ and $|\rho| > \sqrt{\frac{1}{2}\left(\frac{1}{\gamma_y^2} - 1\right)}$.

The sampling designs proportional to the total of the non-observed values in the sample is as follows:

$$P_I(s) \propto N\bar{x} - n\bar{x}_s$$

and the following ones:

$$P_{II}(s) \propto (N - 1)v(x) - (n - 1)v_s(x),$$

$$P_{III}(s) \propto \left(\frac{N}{n} - 1\right)v_\#(x) - n(\bar{x}_s - \bar{x})^2,$$

$$P_{IV}(s) \propto (\bar{x}_s - \bar{x})^2.$$

Wywiał (2000) derived the inclusion probabilities for the above sampling designs as well as the sequences of the conditional probabilities for implementing the sampling designs. The approximate variance of the Horvitz–Thompson's estimator under the defined sampling designs is approximately equal to the variance of the simple random sample mean drawn without replacement.

2.4   Sampling Designs Proportional to the Generalized Variance

The regression sampling strategy, defined as a pair of the ordinary regression estimator and the Singh–Srivastava's sampling design in the Sect. 2.3, will be generalized
into the case of a multidimensional auxiliary variable. First, let us introduce some
notation. The vector $\mathbf{y} = [y_1 \ldots y_N]^T$ consists of all the values of a variable under
study. Let $\mathbf{X} = [x_{kj}]$ be the matrix of the dimensions $N \times m$. The matrix $\mathbf{X}$ consists
of all the values of an m-dimensional auxiliary variable. The element x_{kj} is the kth
value $(k = 1, \ldots, N)$ of the jth auxiliary variable $(j = 1, \ldots, m \geq 1)$. Let $\mathbf{J}_N$ be
the column vector of the dimensions $N \times 1$. Each element of the vector $\mathbf{J}_N$ is equal
to one. Note the following:

$$\bar{\mathbf{x}} = \frac{1}{N}\mathbf{J}_N^T\mathbf{X}, \quad \mathbf{V}(x) = [v(x_i, x_j)] = \frac{1}{N}(\mathbf{X} - \mathbf{J}_N\bar{\mathbf{x}})^T\mathbf{X} - \mathbf{J}_N\bar{\mathbf{x}}),$$

$$\mathbf{v}(x, y) = [v(x_i, y)] = \frac{1}{N}(\mathbf{X} - \mathbf{J}_N\bar{\mathbf{x}})^T(\mathbf{y} - \mathbf{J}_N\bar{y}).$$

The row vector $\bar{\mathbf{x}}$ consists of the population means of the auxiliary variables. The
population variance–covariance matrix of the auxiliary variables has been denoted
by $\mathbf{V}(x)$. The vector $\mathbf{v}(x, y)$ consists of the population covariances of the auxiliary
variables and the variable under study. Let $\mathbf{y}_s = [y_{k_1} \ldots y_{k_n}]^T$ be the vector of values
of the variable under study observed in the sample s of size n. Similarly, the matrix

$$\mathbf{X}_s = \begin{bmatrix} x_{k_1 j} \ldots x_{k_1 m} \\ x_{k_2 j} \ldots x_{k_2 m} \\ \ldots\ldots\ldots\ldots \\ x_{k_n j} \ldots x_{k_n m} \end{bmatrix}$$

consists of the values of the auxiliary variables observed in the sample s. Moreover,
the sample variance–covariance matrix can be rewritten as follows:

$$\bar{\mathbf{x}}_s = \mathbf{J}_n^T\mathbf{X}_s, \quad \mathbf{V}_s(x) = [v_s(x_i, x_j)] = \frac{1}{n}(\mathbf{X}_s - \mathbf{J}_n\bar{\mathbf{x}}_s)^T(\mathbf{X}_s - \mathbf{J}_n\bar{\mathbf{x}}_s),$$

$$\mathbf{v}_s(x, y) = [v_s(x_i, y)] = \frac{1}{n}(\mathbf{X}_s - \mathbf{J}_n\bar{\mathbf{x}}_s)^T(\mathbf{y}_s - \mathbf{J}_n\bar{y}_s).$$

Therefore, $\mathbf{V}_s(x) = [v_s(x_i, x_j)]$ is the sample variance–covariance matrix between
the auxiliary variables and $\mathbf{v}_s(x, y)$ is the column vector of the sample covariances
between the auxiliary variables and the variable under study. The generalized variance
is the determinant of the variance covariance matrix. On the basis of the results
by Wywiał (1999a,b) the following sampling strategies dependent on generalized
variance are shown.

Sampling Strategy 1

The first sampling design proportional to the sample generalized variance is

$$P_V(s) = c_1 \frac{\det \mathbf{V}_s(x)}{\det \mathbf{V}(x)} \tag{2.7}$$

where

$$c_1 = \left(\frac{N - m - 1}{n - m - 1} \right)^{-1} \left(\frac{n}{N} \right)^{m+1}$$

When $m = 1$, the sampling design $P_V(s)$ is reduced to the sampling design $P_{SS}(s)$ of Singh and Srivastava (1980), which was considered in the Sect. 2.3.

Let $\mathbf{X}(k_1, \ldots, k_h)$ be a sub-matrix obtained through dropping the rows of numbers $k_1, \ldots, k_h$ in the matrix $\mathbf{X}$. Moreover, let

$$\mathbf{F}(k_1, \ldots, k_h) = \mathbf{X}(k_1, \ldots, k_h) - \bar{\mathbf{x}}(k_1, \ldots, k_h)\mathbf{J}_{N-h}$$

$\mathbf{J}_{N-h}$ is the column vector with all its $(N - h)$ elements equal to one and

$$\bar{\mathbf{x}}(k_1, \ldots, k_h) = \frac{1}{N - h}\mathbf{J}_{N-h}\mathbf{X}(k_1, \ldots, k_h)$$

The inclusion probabilities of are as follows:

$$\pi_k = 1 - \frac{\binom{N-m-2}{n-m-1}(N - 1)}{\binom{N-m-1}{n-m-1}N^{m+1} \det \mathbf{V}(\mathbf{X})}|\mathbf{F}(k)^T\mathbf{F}(k)|, \, k = 1, \ldots, N$$

$$\pi_{hk} = 1 - \frac{1}{\binom{N-m-1}{n-m-1}N^{m+1} \det \mathbf{V}(\mathbf{X})} \cdot$$

$$\cdot \left\{ \binom{N - m - 2}{n - m - 1}(N - 1)\left[|\mathbf{F}(h)^T\mathbf{F}(h)| + |\mathbf{F}(k)\mathbf{F}^T(k)|\right] - \right.$$

$$\left. \binom{N - m - 3}{n - m - 1}(N - 2)|\mathbf{F}(h, k)\mathbf{F}^T(h, k)| \right\}$$

$h = 1, \ldots, N, k = 1, \ldots, N, k \neq h$.

Let s_{m+1} be the subset of the sample s. The size of the subset s_{m+1} is equal to $m + 1 < n$. Let

$$q_1(s_{m+1}) = \det[\mathbf{J}_{m+1} \, \mathbf{X}_{s_{m+1}}]$$

where $\mathbf{X}_{s_{m+1}} = \begin{bmatrix} x_{k_1 j} \dots x_{k_1 m} \\ \dots \dots \\ x_{k_m 1} \dots x_{k_m m} \\ x_{k_{m+1} 1} \dots x_{k_{m+1} m} \end{bmatrix}$

or

$$q_1(s_{m+1}) = \overset{2}{\det}[\mathbf{X}_{s_m} - \mathbf{x}_{m+1*} \, \mathbf{J}_m]$$

where $\mathbf{x}_{r*} = \lfloor x_{k_r 1} \dots x_{k_r m} \rfloor$ is the $m + 1$th row of the matrix $\mathbf{x}_{s_{m+1}}$. After dropping the row $\mathbf{x}_{m+1*}$ in the matrix $\mathbf{x}_{s_{m+1}}$, we obtain the matrix $\mathbf{x}_{s_m}$.

Let us note that $q_1(s_{m+1})$ is the m-dimensional measure (volume) of the parallelotop spanned by the vectors with their origins at the same point $\mathbf{x}_{m+1*}$ and the end points $\mathbf{x}_{1*}, \dots, \mathbf{x}_{m*}$, (see e.g. Borsuk 1969). From another point of view, $q_1(s_{m+1})$ is proportionate to the m-dimensional volume of the simplex spanned by the points $\mathbf{x}_{1*}, \dots, \mathbf{x}_{m+1*}$.

The sampling scheme (implementing the sampling design $P_V(s)$) consists of the two following steps:

Step 1: Select $(m + 1)$ units $s_{m+1} = \{k_1, k_2, \dots, k_{m+1}\}$ with their probability of joint selection being proportional to $q_1(s_{m+1})$.
Step 2: Select $(n - m - 1)$ units from the remaining units of the population by a simple random sampling without replacement.

The well-known multiple regression estimator of the population mean $\overline{y}$ is as follows:

$$\overline{y}_{Reg,s} = \overline{y}_s - (\overline{\mathbf{x}}_s - \overline{\mathbf{x}})\mathbf{B}_s \tag{2.8}$$

where

$$\mathbf{B}_s = \mathbf{V}_s^{-1}(x)\mathbf{v}_s(x, y)$$

Let

$$\mathbf{A}_s = \begin{bmatrix} \overline{y}_s - \overline{y} & \overline{\mathbf{x}}_s - \overline{\mathbf{x}} \\ \mathbf{v}_s(x, y) & \mathbf{V}_s(x) \end{bmatrix}$$

The well-known property of the determinant of a block matrix lets us rewrite the estimator $\overline{y}_{Reg,s}$ in the following way:

$$\overline{y}_{Reg,s} = \overline{y} + \frac{\det \mathbf{A}_s}{\det \mathbf{V}_s(x)} \tag{2.9}$$

$(\overline{y}_{Reg,s}, P_V(s))$ is the unbiased strategy of the population mean $\overline{y}$. When the sample size $n \to \infty$, the population size $N \to \infty$ and $N - n \to \infty$, then

$$V(\bar{y}_{Reg,s}, P_V(s)) \approx \frac{1}{n}(v(y) - \mathbf{v}^T(x, y)\mathbf{V}^{-1}(x)\mathbf{v}(x, y)) \tag{2.10}$$

or

$$V(\bar{y}_{Reg,s}, P_V(s)) \approx \frac{1}{n}v(y)(1 - r_w^2) \tag{2.11}$$

where

$$r_w = \sqrt{\mathbf{r}^T \mathbf{R}^{-1}(x)\mathbf{r}}$$

is the multiple correlation coefficient between the auxiliary variables and the variable under study. The matrix $\mathbf{R}(x)$ is the correlation matrix of auxiliary variables and $\mathbf{r}^T = [r_{y1} \ldots r_{ym}]$, where r_{yj} is the correlation coefficient between the jth auxiliary variable and the variable under study. Hence, in the asymptotic case, the precision of the strategy $(\bar{y}_{Reg,s}, P_V(s))$ increases when the value of the multiple correlation coefficient r_w also increases.

The unbiased estimator of the variance $V(\bar{y}_{Reg,s}, P_V(s))$ is as follows:

$$\hat{V}_{1s} = \bar{y}_{Reg,s}^2 - \frac{N^{m-1}\prod_{h=1}^{m}(n-h)\det\mathbf{V}(x)}{n^{m+1}\prod_{h=1}^{m}(N-h)\det\mathbf{V}_s(x)}\left[\sum_{i\in s}y_i^2 + \frac{N-1}{n-1}\sum_{i\ne j\in s}y_iy_j\right] \tag{2.12}$$

Sampling Strategy 2

The next sampling design is proportional to the determinant $\det\mathbf{V}_{\#s}(x)$ where

$$\mathbf{V}_{\#s}(x) = \frac{1}{n}(\mathbf{X}_s - \mathbf{J}_n\bar{\mathbf{x}})^T(\mathbf{X}_s - \mathbf{J}_n\bar{\mathbf{x}})$$

is the sample variance–covariance matrix dependent on the population means of the auxiliary variables. Let us consider the following sampling design:

$$P_{VI}(s) = c_2\frac{\det\mathbf{V}_{\#s}(x)}{\det\mathbf{V}(x)} \tag{2.13}$$

where $c_2 = (\frac{n}{N})^m \left(\dfrac{N-m}{n-m}\right)^{-1}$

When $m = 1$, the defined sampling design is reduced to the following sampling design, (see Singh and Srivastava 1980):

$$P_{VI}(s) = \frac{v_{\#s}(x)}{\binom{N}{n}v_{\#}(x)}, \quad s \in \mathbf{S}$$

where $v_{\#s}(x) = \frac{n-1}{n}v_s(x)$, $v_{\#} = \frac{N-1}{N}v(x)$.

Let $\mathbf{X}_{\#}(k_1, \ldots k_r)$ be a sub-matrix obtained through dropping the rows of numbers $k_1, \ldots, k_r$ from the matrix $\mathbf{X} - \mathbf{J}_N \bar{\mathbf{x}}$. The inclusion probabilities of the first and second orders are as follows:

$$\pi_k = 1 - \frac{N-n}{N-m} \frac{|\mathbf{X}_{\#}(k)\mathbf{X}_{\#}^T(k)|}{N^m |\mathbf{V}(x)|}, \quad k = 1, \ldots, N$$

$$\pi_{hk} = 1 - \frac{N-n}{(N-m)N^m |\mathbf{V}(x)|} \cdot$$
$$\left(|\mathbf{X}_{\#}(h)\mathbf{X}_{\#}^T(h)| + |\mathbf{X}_{\#}(k)\mathbf{X}_{\#}^T(k)| - \frac{N-n-1}{N-m-1} |\mathbf{X}_{\#}(h,k)\mathbf{X}_{\#}^T(h,k)| \right)$$

$h = 1, \ldots, N, k = 1, \ldots, N, k \neq h$.

Let s_m be a subset of the sample s and $m < n$. Let us define the following:

$$q_2(s_m) = \det^2 \begin{bmatrix} 1 & \bar{\mathbf{x}} \\ \mathbf{J}_m & \mathbf{X}_{s_m} \end{bmatrix}$$

From the geometrical point of view, $q_2(s_m)$ is the m-dimensional measure (volume) of the parallelotop spanned by the vectors with their origins at the same point $\bar{\mathbf{x}}$ and the end points that determine the rows of the matrix $\mathbf{X}_{s_m}$, see, e.g. Anderson (1958) or Borsuk (1969). The above expression can be transformed into the following one:

$$q_2(s_m) = \det^2(\mathbf{X}_{s_m} - \mathbf{J}_m \bar{\mathbf{x}}) \tag{2.14}$$

The following sampling scheme implements the sampling design $P_{VI}(s)$:

Step 1: Select m-units $s_m = \{k_1, \ldots, k_m\}$ with their probability of joint selection being proportional to $q_2(s_m)$.

Step 2: Select $(n - m)$ units from the remaining units of the population by simple random sampling without replacement.

When $m = 1$, the introduced sampling scheme is reduced to the sampling scheme proposed by Singh and Srivastava (1980). In this case, $q_2(s_1) = q_2(k_1) = (x_{k_1} - \bar{x})^2$, $k_1 = 1, \ldots, N$.

Let us consider the following estimator

$$\bar{y}_{\#s} = \frac{n(N-m)}{N(n-m)} [\bar{y}_s - (\bar{\mathbf{x}}_s - \bar{\mathbf{x}})\mathbf{B}_{\#s}] \tag{2.15}$$

where

$$\mathbf{B}_{\#s} = \mathbf{V}_{\#s}(x)^{-1} \mathbf{v}_{\#s}(x, y), \quad \mathbf{v}_{\#s}(x, y) = \frac{1}{n}(\mathbf{X}_{s_m} - \mathbf{J}_m \bar{\mathbf{x}})^T \mathbf{y}_s$$

The statistic $\bar{y}_{\#s}$ can be transformed into the following one:

$$\bar{y}_{\#s} = \frac{n(N-m)\det \mathbf{A}_{\#s}}{N(n-m)\det \mathbf{V}_{\#s}(x)} \qquad (2.16)$$

where

$$\mathbf{A}_{\#s} = \begin{bmatrix} \bar{\mathbf{y}}_s & \bar{\mathbf{x}}_s - \bar{\mathbf{x}} \\ \mathbf{v}_{\#s}(x,y) & \mathbf{V}_{\#s}(x) \end{bmatrix}$$

or

$$\mathbf{A}_{\#s} = \frac{1}{n}\begin{bmatrix} \mathbf{J}_n^T \\ (\mathbf{X}_{s_m} - \mathbf{J}_m\bar{\mathbf{x}})^T \end{bmatrix}[\mathbf{y}_s \ (\mathbf{X}_{s_m} - \mathbf{J}_m\bar{\mathbf{x}})]$$

In the case when $m = 1$, the statistic $\bar{y}_{\#s}$ is reduced to the estimator proposed by Singh and Srivastava (1980). $(\bar{y}_{\#s}, P_{VI}(s))$ is the unbiased strategy of the population mean $\bar{y}$. The approximate variance of the strategy $(\bar{y}_{\#s}, P_{VI}(s))$ is expressed by the Eqs. (2.10) or (2.11). The unbiased estimator of the variance of the sampling strategy $(\bar{y}_{\#s}, P_{VI}(s))$ is as follows:

$$\hat{V}_{2s} = \bar{y}_{\#s}^2 - \frac{N^{m-3}\prod_{h=1}^{m}(n-h+1)}{n^{m-1}\prod_{h=2}^{m}(N-h+1)}\frac{\det \mathbf{V}_\#}{\det \mathbf{V}_{\#s}}\left[\sum_{i\in s}y_i^2 + \frac{N-1}{n-1}\sum_{i\neq j\in s}y_iy_j\right]. \qquad (2.17)$$

Sampling Strategy 3

Let us assume that a population U is divided into disjoint and non-empty clusters U_g, $g = 1, \ldots, G$, and $U = \cup_{g=1}^{G}U_g$. The size of the cluster U_g is denoted by $N_g > 1$ and $N = \cup_{g=1}^{G}N_g$. Let the clusters be the first stage units. The sample $s_I = (g_1, \ldots, g_i, \ldots, g_n)$ consisting of clusters, is selected at the first stage. The size of the sample s_I is equal to n. At the second stage the simple random samples $s_{g_1/s_I}, \ldots, s_{g_i/s_I}, \ldots, s_{g_n/s_I}$ where $g_i \in s_I$ are drawn without replacement from the clusters $U_{g_1}, \ldots, U_{g_i}, \ldots, U_{g_n}$, respectively, selected on the first stage. The size of the sample s_{g_i} is denoted by $1 < m_{g_i} \leq N_{g_i}$. The two-stage sample will be denoted by $s = \{s_I, s_{g_1/s_I}, \ldots, s_{g_i/s_i}, \ldots, s_{g_n/s_I}\}$. Let a multivariate auxiliary variable of the dimension m be observed on the all first-stage units. So, it means that a value x_{gj} of a jth auxiliary variable is attached to the gth first-stage unit, $g = 1, \ldots, G$. The first-stage sampling design is proportional to the sample generalized variance of the auxiliary variables given by the expression (2.7) where the population size N has to be replaced with the number of the first-stage unit G. So, the sampling design of the two-stage sample is as follows:

$$P_{VII}(s) = P_V(s_I) \prod_{g \in s_I} \binom{N_g}{n_g}^{-1} \tag{2.18}$$

where the sampling design $P_V(s_I)$ is determined by the Eq. (2.7) where the population size N has to be replaced with the number of the first-stage unit G. Let us introduce the following notation:

$$z_g = \sum_{i \in U_g} y_i, \quad \bar{z} = \frac{1}{G} \sum_{g=1}^{G} z_g, \quad v_z = \frac{1}{G-1} \sum_{g=1}^{G} (z_g - \bar{z})^2,$$

$$\bar{y}_{U_g} = \frac{1}{N_g} z_g, \quad v_{U_g} = \frac{1}{N_g - 1} \sum_{i \in U_g} (y_i - \bar{y}_{U_g})^2, \quad z_{s_g/s_I} = \sum_{i \in s_g/s_I} y_i,$$

$$\bar{y}_{s_g/s_I} = \frac{1}{n_g} z_{s_g/s_I}, \quad \tilde{z}_{s_g/s_I} = N_g \bar{y}_{s_g/s_I}, \quad v_{s_g/s_I} = \frac{1}{n_g - 1} \sum_{i \in s_g/s_I} (y_i - \bar{y}_{s_g/s_I})^2.$$

The statistics $\tilde{z}_{s_g/s_I}$ and v_{s_g/s_I} are unbiased estimators of the cluster total z_g and the cluster variance v_{U_g}, respectively.

Let us consider the following estimator of the population mean $\bar{y}$ where

$$\bar{y}_{reg2s} = \frac{G}{N}(\bar{z}_s - (\bar{\mathbf{x}}_{s_I} - \bar{\mathbf{x}}\mathbf{B}_s)) \tag{2.19}$$

where

$$\bar{z}_s = \frac{1}{n} \sum_{g \in s_I} \tilde{z}_{s_g/s_I}, \tag{2.20}$$

$$\bar{\mathbf{x}}_{s_I} = [\bar{x}_{s_I,1} \dots \bar{x}_{s_I,j} \dots \bar{x}_{s_I,m}], \quad \bar{x}_{s_I,j} = \frac{1}{n} \sum_{g \in s_I} x_{gj}, \quad j = 1, \dots, m.$$

$$\mathbf{B}_s = \mathbf{v}_{s_I}^{-1} \mathbf{w}_s, \quad \mathbf{v}_{s_I} = [v_{s_I}(x_i, x_j)],$$

$$v_{s_I}(x_i, x_j) = \frac{1}{n-1} \sum_{g \in s_I} (x_{gj} - \bar{x}_{s_I i})(x_{gj} - \bar{x}_{s_I,j}),$$

$$\mathbf{w}_s^T = [v_s(z, x_1), \dots, v_s(z, x_m)], \quad v_s(z, x_g) = \frac{1}{n-1} \sum_{g \in s_I} (\tilde{z}_{s_g/s_I} - \bar{z}_s)(x_{gj} - \bar{x}_{s_I,j}).$$

The statistic $\bar{y}_{reg2s}$ is the unbiased estimator for the population mean and its variance is

$$V(\overline{y}_{reg2s}) \approx \frac{G^2}{n^2 N^2} E\left(\sum_{g \in s_I} v(\tilde{z}_{s_{g/s_I}}) + \overline{x}_{s_I} V_{s_I}^{-1} X_{s_I}^T diag(v_{s/s_I}(\tilde{z}_{s/s_I})) X_{s_I} V_{s_I}^{-1} \overline{x}_{s_I}^T + \right.$$

$$2\overline{x}_{s_I} V_{s_I}^{-1} X_{s_I}^T V_{s/s_I}(\tilde{z}_{s/s_I})(1 + \overline{x}_{s_I} V_{s_I}^{-1} \overline{x}_{s_I}^T) +$$

$$\left. \overline{x}_{s_I} V_{s_I}^{-1} X_{s_I}^T v_{s/s_I}(\tilde{z}_{s/s_I})(2 + \overline{x}_{s_I} V_{s_I}^{-1} \overline{x}_{s_I}^T) \sum_{g \in s_I} v(\tilde{z}_{s_{g/s_I}}) \right) + \frac{1}{n} v_z(1 - r_{zx}^2) \quad (2.21)$$

where x_{s_I} is the $n \times m$ matrix of the auxiliary variables observed in the sample s_I; the column vector $\tilde{z}_{s/s_I}$:

$$\tilde{z}_{s/s_I} = \begin{bmatrix} \tilde{z}_{s_1/s_I} \\ \cdots\cdots \\ \tilde{z}_{s_n/s_I} \end{bmatrix}$$

consists of the statistics $\tilde{z}_{s_{g/s_I}}$, $g = 1, ..., n$;
the column vector $v_{s/s_I}(\tilde{z}_{s/s_I})$:

$$v_{s/s_I}(\tilde{z}_{s/s_I}) = \begin{bmatrix} v(\tilde{z}_{s_{g/s_I}}) \\ \cdots\cdots \\ v(\tilde{z}_{s_{n/s_I}}) \end{bmatrix}$$

consists of the following conditional variance (under the fixed sample s_I) of the statistics $\tilde{z}_{s_{g/s_I}}$, $g = 1, ..., G$:

$$v(\tilde{z}_{s_{g/s_I}}) = \frac{N_g(N_g - n_g)}{n_g} v_{U_g};$$

r_{zx} is the multiple correlation coefficient between the auxiliary variables and the variable z.

The following statistic is the unbiased estimator of the variance $V(\overline{y}_{reg2s})$:

$$V_{3,s} = \overline{z}_s^2 - \frac{N^{m-1} \prod_{h=1}^{m}(n-h)}{N^{m+1} \prod_{h=1}^{m}(N-h)} \frac{\det V(x)}{\det V_{s_I}(x)} \left[\sum_{g \in s_I} \check{z}_{s_{g/s_I}} + \frac{N-1}{n-1} \sum_{g \neq t \in s_I} \tilde{z}_{s_{g/s_I}} \tilde{z}_{s_{t/s_I}} \right]$$
$$(2.22)$$

where

$$\check{z}_{s_{g/s_I}} = \frac{n_g(N_g - 1)}{N_g(n_g - 1)} \left(\tilde{z}_{s_{g/s_I}}^2 - \frac{N_g^2}{n_g} \frac{N_g - n_g}{N_g - 1} M_{s_{g/s_I}} \right), \quad M_{s_{g/s_I}} = \frac{1}{n_g} \sum_{i \in s_{g/s_I}} y_i^2.$$

Based on the simulation analysis, Gamrot and Wywiał (2002) compared the accuracy of the estimator $\overline{y}_{reg2s}$ with the weighted mean of the variable under study observed in the two-stage sample proposed by Rao et al. (1962) or by Hartley and Rao (1962). In the first stage, the sample was selected by means of the well-known sampling designs proportional to the value of an auxiliary variable. In the second stage, the simple random samples were selected without replacement from the previously drawn clusters. The computer simulation analysis was based on the empirical set of data available in the book by Särndal et al. (1992). In general, the analysis leads to the conclusion that the accuracies of both sampling strategies are comparable.

The sampling designs under consideration can be useful in several types of scientific research supported by statistical analysis. For instance (see, e.g. Wywiał 2013), the sampling design proportional to the generalized variance of a two-dimensional auxiliary variable (values of the auxiliary variables are treated as geographic coordinates) can be useful in the sampling of populations considered in regional economic research, agriculture, geodesy, ecology, and so on. The sampling design proportional to the generalized variance of a three-dimensional auxiliary variable can be applied in geological research or ecology, as well. Moreover, the sampling design proportional to the trace of the sample variance–covariance matrix leads to samples in which population objects are distant from each other. Let us note that in this case we can consider the objects in more than three dimensions. For instance, census data on households can be treated as observations of a multidimensional auxiliary variable. The sample may be selected proportionally to the generalized variance of standardized observations of variables: e.g. the age of head of household, the apartment area, the number of children, the level of rent, and the household income over the last year. Thus, we have a five-dimensional space for observations of the auxiliary variable. In this case, the sampling design leads to the selection of households for which observations of auxiliary variables are not linearly dependent in the sense of the determinant of the matrix $\mathbf{X}_s^T \mathbf{X}_s$. Hence, this can lead to improving the accuracy of the estimation of the linear regression model parameters between auxiliary variables and variables under study on the basis of the well-known method of least squares. Finally, we can note that it is possible to find several more examples, which let us at least modify the sampling strategies under considerations.

References

Anderson, T. W. (1958). *An Introduction to Multivariate Statistical Analysis*. New York: Wiley.

Bhushan, S., Singh, R. K., & Katara, S. (2009). Improved estimation under Midzuno–Lahiri–Sen type sampling design. *Journal of Reliability and Statistical Studies, 2*(2), 59–66.

Borsuk, K. (1969). *Multidimensional Analytic Geometry*. Warsaw: PWN.

Brewer, K. R. W., & Hanif, M. (1983). *Sampling with Unequal Probabilities*. New York-Heidelberg-Berlin: Springer.

Chaudhuri, A., & Arnab, R. (1981). On non-negative variance-estimation. *Metrika, 28*, 1–12.

Chaudhuri, A., & Stenger, H. (2005). *Survey Sampling. Theory and Methods* (2nd ed.). Boca Raton-London-New York-Singapore: Chapman & Hall CRC.

Gamrot, W., & WywiałJ, L. (2002). Comparison of the accuracy of some two-stage sampling schemes by means of computer simulation. *Zeszyty Naukowe Akademii Ekonomicznej w Katowicach, 21*, 11–28.

Hartley, H. O., & Rao, J. N. K. (1962). Sampling with unequal probabilities and without replacement. *Annals of Mathematical Statistics, 33*, 350–374.

Horvitz, D. G. & Thompson, D. J. (1952) A generalization of the sampling without replacement from finite universe. *Journal of the American Statistical Association, 47*, 663–685.

Lahiri, D. B. (1951). A method for sample selection providing unbiased ratio estimator. *Bulletin of the International Statistical Institute, 33*(2), 133–140.

Midzuno, H. (1952). On the sampling system with probability proportional to sum of sizes. *Annals of the Institute of Matematics and Statistisc, 3,* 99–107.

Rao, J. N. K., Hartley, H. O., & Cochran, W. G. (1962). On a simple procedure of unequal probability sampling without replacement. *Journal of the Royal Statistical Association B, 24*(2), 482–491.

Rao, T. J. (1966). On the variance of the ratio estimator for Midzuno–Sen sampling design. *Metrika, 10,* 89–91.

Rao, T. J. (1977). Estimating the variance of the ratio estimator for the Midzuno–Sen sampling scheme. *Metrika, 24,* 203–208.

Sampford, M. R. (1967). On sampling without replacement with unequal probabilities of selection. *Biometrika, 54,* 499–513.

Särndal, C. E., Swensson, B., & Wretman, J. (1992). *Model Assisted Survey Sampling.* New York-Berlin-Heidelberg-London-Paris-Tokyo-Hong Kong- Barcelona-Budapest: Springer.

Sen, A. R. (1953). On the estimate of variance in sampling with varying probabilities. *Journal of the Indian Society of Agicultural Statistics, 5*(2), 119–127.

Singh, P., Srivastava, A. K. (1980). Sampling schemes providing unbiased regression estimators. *Biometrika, 67*(1), 205–209.

Srivenkataramana, T. (2002). Location-shifts for improved estimation under Midzuno–Sen sampling scheme. *Journal of Statistical Planning and Inference, 102,* 179–187.

Tillé, Y. (2006). *Sampling Algorithms.* Berlin: Springer.

Walsh, J. E. (1970). Generalization of ratio estimator for population total. *Sankhya, A, 32,* 99–106.

Wywiał, J. L. (1999a). Generalisation of Singh and Srivastava's schemes providing unbiased regression estimatiors. *Statistics in Transition, 4*(2), 259–281.

Wywiał, J. L. (1999b). Sampling designs dependent on the sample generalized variance of auxiliary variables. *Journal of the Indian Statistical Association, 37,* 73–87.

Wywiał, J. L. (2000). On precision of Horvitz–Thompson strategies. *Statistics in Transition, 4*(5), 779–798.

Wywiał, J. L. (2003). *Some Contributions to Multivariate Methods in Survey Sampling.* Katowice University of Economics, Katowice. https://www.ue.katowice.pl/fileadmin/user_upload/wydawnictwo/Darmowe_E-Booki/Wywial_Some_Contributions_To_Multivariate_Methods_In_Survey_Sampling.pdf.

Wywiał, J. L. (2013). On space sampling. In B. Suchecki (Ed.), *Spatial Econometrics and Regional Economics Analysis* (no. 292, pp. 21–35). Acta Universitatis Lodziensis. Folia Oeconomica.

Chapter 3
Sampling Designs Based on Order Statistics of an Auxiliary Variable

In this chapter, the properties of sampling designs proportional to the sample order statistic (quantile) as well as functions of two- or three-order statistics are considered in detail. Conditional sampling designs are also taken into account and exact expressions for inclusion probabilities are presented. The sampling schemes for these sampling designs are also proposed, some of which are generalized to the case of continuous-type sampling designs in the last chapter of the book.

3.1 Basic Properties of Order Statistics

We will consider sampling strategies based on sampling designs dependent on functions of order statistics of an auxiliary variable. The distribution of the order statistics is evaluated on the basis of a simple random sample of size n drawn without replacement. Conditional sampling designs are also defined. Moreover, some estimation strategies of the population means are also considered.

First, the probability distribution of the first-order statistic from a simple random sample mean drawn without replacement will be explained. Next, the joint distributions of the second- or third-order statistics will also be considered.

Let $x_k \leq x_{k+1}$ for $k = 1, ..., N - 1$. The sequence of the order statistics of observations of an auxiliary variable in the sample s will be denoted by $(X_{(j)})$. Let $G(r, i) = \{s : X_{(r)} = x_i\}$ be the set of all samples whose rth-order statistic of the auxiliary variable is equal to x_i, where $r \leq i \leq N - n + r$. The size of the set $G(r, i)$ is denoted by $g(r, i) = Card(G(r, i))$ where

$$g(r, i) = \binom{i - 1}{r - 1}\binom{N - i}{n - r}. \qquad (3.1)$$

J. L. Wywiał, *Sampling Designs Dependent on Sample Parameters of Auxiliary Variables*, SpringerBriefs in Statistics, https://doi.org/10.1007/978-3-662-63413-4_3

The equality $\bigcup_{i=r}^{N-n+r} G(r, i) = \mathbf{S}$ leads to the following

$$Card \left(\bigcup_{i=r}^{N-n+r} G(r, i) \right) = \sum_{i=r}^{N-n+r} Card(G(r, i)) = \sum_{i=r}^{N-n+r} g(r, i) = Card(\mathbf{S}) = \binom{N}{n}.$$

Wilks (1962), pp. 243–244, shows that the probability that the rth-order statistic is equal to x_i is as follows (see Guenther 1975):

$$P \left(X_{(r)} = x_i \right) = P \left(s \in G(r, i) \right) = \frac{g(r, i)}{\binom{N}{n}}. \tag{3.2}$$

$$E \left(X_{(r)} \right) = \sum_{i=r}^{N-n+r} x_i P \left(X_{(r)} = x_i \right) = \frac{1}{\binom{N}{n}} \sum_{i=r}^{N-n+r} x_i g(r, i). \tag{3.3}$$

The sample quantile of order $\alpha \in (0; 1)$ can be defined by the equation:

$$Q_\alpha = X_{(r)} \tag{3.4}$$

where $r = [n\alpha] + 1$ is the integer part of the value $n\alpha$, $r = 1, 2, ..., n$. Hence, $X_{(r)} = Q_\alpha$ for $\frac{r-1}{n} \leq \alpha < \frac{r}{n}$.

Let us note that $s = \{s_1, i, s_2\}$ where $s_1 = \{i_1, ..., i_{r-1}\}$, $s_2 = \{i_{r+1}, ..., i_n\}$, $i_j < i$ for $j = 1, ..., r - 1$, $i_r = i$ and $i_j > i$ for $j = r + 1, ..., n$.

The properties of the joint probability distribution of two-order statistics are as follows. Let $G(r_1, r_2, i, j) = \{s X_{(r_1)} = x_i, X_{(r_2)} = x_j\}$, $r_1 = 1, .., n - 1$; $r_2 = 2, ..., n$, $r_1 < r_2$ be the set of all samples whose r_1th- and r_2th-order statistics of an auxiliary variable are equal to x_i and x_j, respectively, where $r_1 \leq i < j \leq N - n + r_2$. Therefore, we have

$$\bigcup_{i=r_1}^{N-n+r_1} \bigcup_{j=i+r_2-r_1}^{N-n+r_2} G(r_1, r_2, i, j) = \mathbf{S}. \tag{3.5}$$

The size of the set $G(r_1, r_2, i, j)$ is denoted by $g(r_1, r_2, i, j) = Card(G(r_1, r_2, i, j))$ and

$$g(r_1, r_2, i, j) = \binom{i-1}{r_1-1}\binom{j-i-1}{r_2-r_1-1}\binom{N-j}{n-r_2}. \tag{3.6}$$

Hence, the joint distribution of the statistics $(X_{(r_1)}, X_{(r_2)})$ is as follows:

$$P \left(X_{(r_1)} = x_i, X_{(r_2)} = x_j \right) = P \left(s \in G(r_1, r_2, i, j) \right) = \frac{g(r_1, r_2, i, j)}{\binom{N}{n}} \tag{3.7}$$

where $i = r_1, ..., N - n + r_1$, $j = r_2, ..., N - n + r_2$, $j > i$.

The basic properties of the joint probability distribution of the three order statistics denoted by $(X_{(r_1)}, X_{(r_2)}, X_{(r_3)})$ where $1 \leq r_1 < r_2 < r_3 \leq n$, are as follows. Let $G(r_1, r_2, r_3; i_1, i_2, i_3) = \{s : X_{(r_1)} = x_{i_1}, X_{(r_2)} = x_{i_2}, X_{(r_3)} = x_{i_3}\}$, be the set of all samples whose r_1th-, r_2th-, and r_3th-order statistics of the auxiliary variable are equal to x_{i_1}, x_{i_2} and x_{i_3}, respectively, where $r_1 \leq i_1 < i_2 < i_3 \leq N - n + r_3$. So, we have

$$\bigcup_{i_1=r_1}^{N-n+r_1} \bigcup_{i_2=i_1+r_2-r_1}^{N-n+r_2} \bigcup_{i_3=i_1+r_3-r_2}^{N-n+r_3} G(r_1, r_2, r_3; i_1, i_2, i_3) = \mathbf{S}.$$

The size of the set $G(r_1, r_2, r_3; i_1, i_2, i_3)$ is denoted by $g(r_1, r_2, r_3; i_1, i_2, i_3) = Card(G(r_1, r_2, r_3; i_1, i_2, i_3))$ and

$$g_3 = g(r_1, r_2, r_3; i_1, i_2, i_3) = \binom{i_1 - 1}{r_1 - 1}\binom{i_2 - i_1 - 1}{r_2 - r_1 - 1}\binom{i_3 - i_2 - 1}{r_3 - r_2 - 1}\binom{N - i_3}{n - r_3}$$

$$(3.8)$$

The joint distribution of the statistics $X_{(r_1)}, X_{(r_2)}, X_{(r_3)}$ is

$$P\left(X_{(r_1)} = x_{i_1}, X_{(r_2)} = x_{i_2}, X_{(r_3)} = x_{i_3}\right) = P(s \in G(r_1, r_2, r_3; i_1, i_2, i_3)) =$$

$$\sum_{s \in G(r_1, r_2, r_3; i_1, i_2, i_3)} P(s) = \frac{g(r_1, r_2, r_3; i_1, i_2, i_3)}{\binom{N}{n}} \quad (3.9)$$

where $i_1 = r_1, ..., N - n + r_1, i_2 = i_1 + r_2 - r_1, ..., N - n + r_2, i_3 = i_2 + r_3 - r_2,$
$..., N - n + r_3$.

3.2 Sampling Design Proportional to Function of One-Order Statistic

The previous chapter was presented the sampling design and scheme proposed by Sampford which is of type sampling designs with inclusion probabilities proportional to values of the positive auxiliary variable. There is a lot of such sampling designs like it was mentioned above but still there is no many sampling designs with evaluated simple expression for the inclusion of the second degree. It was thee mine inspiration for looking for the new sampling design for which those probabilities are exactly defined. The sampling designs proposed by Wywiał (2008) are defined as follows:

Definition 3.1 The sampling design proportional to the values $x_i, i = r, ..., N - n + r$, of the order statistic $X_{(r)}$ is as follows:

$$P_r(s) = \frac{x_i}{\binom{N}{n}E\left(X_{(r)}\right)} \quad \text{for } s \in G(r, i) \quad (3.10)$$

where $E\left(X_{(r)}\right)$ is given by the expression (3.3).

Thus, in this case, the probability of selecting a sample s is proportional to the value x_i of the order statistic $X_{(r)}$.

Let $f(x_i)$ be an increasing positive function of a value x_i of the order statistic $X_{(r)}$ of the auxiliary variable. The generalization of the above defined sampling design is as follows.

Definition 3.2 The conditional sampling design proportional to the values $f(x_i)$, $i = u, ..., v \leq N - n + r$, $u \geq r$, of $f(X_{(r)})$ is as follows:

$$P_r(s|x_u \leq X_{(r)} \leq x_v) = P_r(s|u, v) = \frac{f(x_i)}{z_r(u, v)} \qquad (3.11)$$

where $i \in s \in G(r, i),\quad r \leq u \leq i \leq v \leq N - n + r$

$$z_r(u, v) = \sum_{j=u}^{v} f(x_j)g(r, j).$$

The unconditional version of the sampling design $P_r(s|u, v)$ is obtained when $u = r$ and $v = N - n + r$.

Definition 3.2 lets us rewrite the conditional sampling design $P_r(s|u, v)$ in another way. The condition $u \leq i \leq v$ can be replaced by $c_1 \leq x_i \leq c_2$. Thus, the conditional sampling design $P_r(s|u, v)$ can be defined as follows:

$$P_r(s|u, v) = P_r(s|c_1, c_2) = \frac{f(x_i, c_1, c_2)}{\sum_{i=r}^{N-n+r} g(r, i)f(x_i, c_1, c_2)}$$

where

$$f(x_i, c_1, c_2) = \begin{cases} f(x_i) & \text{for } c_1 \leq x_i \leq c_2 \\ 0 & \text{for } x_i < c_1 \text{ or } x_i > c_2 \end{cases}$$

and $f(x_i)$ is non-negative as it was previously assumed.

The sampling design $P_r(s|u, v)$ does not depend on those values of the auxiliary variable which are less than x_u and greater than x_v. This means that the sampling design is useful in the case of the singly left or singly right censored observations of the auxiliary variable in the population.

Particularly, when $u = r$, $v = N - n + r$ and $f(x_i) = x_i$, the sampling design $P_r(s|u, v)$ reduces to that given by definition 3.1. Moreover, let us note that in general the considered concept of the conditional sampling design agrees with the definition of the conditional sampling design introduced by Tillé (1998, 2006).

The above definitions are also valid in the case of non-distinct values of the auxiliary variable, too. In this case, the values of some subsets of observations of the auxiliary variable will be the same in the expressions (3.11)–(3.10).

Let us assume that if $x \leq 0$, $\delta(x) = 0$. When $x > 0$, $\delta(x) = 1$. Let us note that $\delta(x)\delta(x - 1) = \delta(x - 1)$. The following theorems are the straightforward generalizations of those proven by Wywiał (2008).

Theorem 3.1 *The inclusion probabilities of the first order for the conditional sampling design $P(s|u, v)$ are as follows:*

$$\pi_k(r|u, v) = \frac{\delta(r - 1)\delta(k - r)}{z_r(u, v)} \sum_{i=u}^{v} \binom{i - 2}{r - 2}\binom{N - i}{n - r} f(x_i) +$$

$$\frac{\delta(r - 1)\delta(k - r + 1)\delta(N - n + r + 1 - k)}{z_r(u, v)} \sum_{i=max(u,k+1)}^{v} \binom{i - 2}{r - 2}\binom{N - i}{n - r} f(x_i) +$$

$$\frac{\delta(n - r)\delta(k - r))\delta(N - n + r + 1 - k))}{z_r(u, v)} \sum_{i=u}^{min(v,k-1)} \binom{i - 1}{r - 1}\binom{N - i - 1}{n - r - 1} f(x_i) +$$

$$\frac{\delta(k + 1 - r)\delta(N - n + r + 1 - k)}{z_r(u, v)} \binom{k - 1}{r - 1}\binom{N - k}{n - r} f(x_k) +$$

$$\frac{\delta(n - r)\delta(k - N - n + r)}{z_r(u, v)} \sum_{i=u}^{v} \binom{i - 1}{r - 1}\binom{N - i - 1}{n - r - 1} f(x_i), \qquad (3.12)$$

Theorem 3.2 *The inclusion probabilities of the second order for the conditional sampling design $P(s|u, v)$ are as follows: if $k < u$, $t < u$ and $t \neq k$*

$$\pi_{k,t}(r|u, v) = \frac{\delta(r - 2)\delta(v - 2)\delta(u - 2)}{z_r(u, v)} \sum_{i=u}^{v} \binom{i - 3}{r - 3}\binom{N - i}{n - r} f(x_i). \qquad (3.13)$$

If $k > v$, $t > v$ and $t \neq k$,

$$\pi_{k,t}(r|u, v) =$$
$$\frac{\delta(n - r - 1)\delta(N - v - 1)\delta(N - u - 1)}{z_r(u, v)} \sum_{i=u}^{v} \binom{i - 1}{r - 1}\binom{N - i - 2}{n - r - 2} f(x_i).$$
$$(3.14)$$

If $k < u$ and $t > v$ or $t < u$ and $k > v$,

$$\pi_{k,t}(r|u, v) = \frac{\delta(r - 1)\delta(n - r)\delta(u - 1)\delta(N - v)}{z_r(u, v)} \sum_{i=u}^{v} \binom{i - 2}{r - 2}\binom{N - i - 1}{n - r - 1} f(x_i).$$
$$(3.15)$$

If $k < u$ and $u \leq t \leq v$ or $t < u$ and $u \leq k \leq v$,

$$\pi_{k,t}(r|u, v) =$$

$$\frac{\delta(r-1)}{z_r(u,v)}\left(\delta(n-r)\delta(t-u)\delta(t-2)\sum_{i=u}^{t-1}\binom{i-2}{r-2}\binom{N-i-1}{n-r-1}f(x_i)+\right.$$

$$\delta(t-1)\binom{t-2}{r-2}\binom{N-t}{n-r}f(x_t)+$$

$$\left.\delta(r-2)\delta(v-t)\delta(v-2)\delta(t-1)\sum_{i=t+1}^{v}\binom{i-3}{r-3}\binom{N-i}{n-r}f(x_i)\right). \qquad (3.16)$$

If $u \le k \le v$ *and* $t > v$ *or* $u \le t \le v$ *and* $k > v$,

$$\pi_{k,t}(r|u,v) = \frac{\delta(n-r)}{z_r(u,v)}\left(\delta(n-r-1)\delta(k-u)\delta(N-k)\delta(k-1)\delta(N-u-1)\cdot\right.$$

$$\sum_{i=u}^{k-1}\binom{i-1}{r-1}\binom{N-i-2}{n-r-2}f(x_i) + \delta(N-k)\binom{k-1}{r-1}\binom{N-k-1}{n-r-1}f(x_k)+$$

$$\left.\delta(r-1)\delta(v-k)\delta(N-v)\delta(v-1)\delta(N-k-1)\sum_{i=k+1}^{v}\binom{i-2}{r-2}\binom{N-i-1}{n-r-1}f(x_i)\right).$$

$$\qquad (3.17)$$

If $u \le k < t \le v$ *or* $u \le t < k \le v$,

$$\pi_{k,t}(r|u,v) = \frac{\delta(v-u)}{z_r(u,v)}\left(\delta(n-r-1)\delta(k-u)\delta(N-k)\delta(k-1)\delta(N-u-1)\cdot\right.$$

$$\sum_{i=u}^{k-1}\binom{i-1}{r-1}\binom{N-i-2}{n-r-2}f(x_i) + \delta(n-r)\delta(N-k)\binom{k-1}{r-1}\binom{N-k-1}{n-r-1}f(x_k)+$$

$$\delta(r-1)\delta(n-r)\delta(t-k-1)\delta(t-2)\delta(N-k-1)\sum_{i=k+1}^{t-1}\binom{i-2}{r-2}\binom{N-i-1}{n-r-1}f(x_i)+$$

$$\delta(r-1)\delta(t-1)\binom{t-2}{r-2}\binom{N-t}{n-r}x_t+$$

$$\left.\delta(r-2)\delta(v-t)\delta(v-2)\delta(t-1)\sum_{i=t+1}^{v}\binom{i-3}{r-3}\binom{N-i}{n-r}f(x_i)\right). \qquad (3.18)$$

Both the above theorems are also true in the case of non-distinct values of the auxiliary variable. In this case, in the expressions (3.13)–(3.18), the values of some subsets of observations of the auxiliary variable will be the same.

Both the theorems lead to the following particular result. If $r \le u = v \le N - n + r$ then

$$P_r(s|u, u) = \frac{P_r(s)}{P_r(s : f(X_{(r)}) = f(x_u))} = \frac{1}{\binom{u-1}{r-1}\binom{N-u}{n-r}}. \qquad (3.19)$$

In this case, the population element with label u is purposively selected from the population. Thus, this indicates that it is selected with the probability of one. Moreover, two samples are drawn independently from the two strata. The first simple random sample of size $r - 1$ is drawn without replacement from the strata consisting of population elements with the labels from $1, ..., u - 1$, and the second simple random sample of size $n - r$ is drawn without replacement from the strata consisting of population elements with the labels from $u + 1, ..., N$. The inclusions probabilities are as follows:

$$\pi_k^{(r)}(u, u) = \begin{cases} \frac{r-1}{u-1} & \text{for } k < u, \\ 1 & \text{for } k = u, \\ \frac{n-r}{N-u} & \text{for } k > u. \end{cases} \tag{3.20}$$

$$\pi_{k,t}^{(r)}(u, u) = \begin{cases} \delta(r-2)\frac{(r-1)(r-2)}{(u-1)(u-2)} & \text{for } k < u, t < u, k \neq t \\ \delta(r-1)\frac{r-1}{u-1} & \text{for } k < u = t, \\ \delta(n-r)\frac{n-r}{N-u} & \text{for } k = u < t, \\ \delta(n-r-1)\frac{(n-r)(n-r-1)}{(N-u)(N-u-1)} & \text{for } k > u, t > u, t \neq t, \\ \delta(r-1)\delta(n-r)\frac{(r-1)(n-r)}{(u-1)(N-u)} & \text{for } k < u < t. \end{cases} \tag{3.21}$$

The two systems above show the inclusion probabilities of the sample selected from two strata when they are determined by means of the fixed value x_u of the auxiliary variable.

The sampling scheme implementing the conditional sampling design $P_r(s|u, v)$ where $r \leq u \leq v \leq N - n + r$ is as follows. First, the population elements are ordered according to increasing values of the auxiliary variable. Next, the ith element of the population where $i = u, u + 1, ..., v$ and $r = [n\alpha] + 1$, is drawn with the probability

$$P_{*,r}(i|u, v) = \frac{f(x_i)g(r, i)}{\sum_{j=u}^{v} f(x_j)g(r, j)} \tag{3.22}$$

Finally, two simple random samples s_1 and s_2 are drawn without replacement from the sub-populations $U_1 = \{1, ..., i - 1\}$ and $U_2 = \{i + 1, i + 2, ..., N\}$, respectively. The sample s_1 is of the size $r - 1$ and the sample s_2 is of the size $n - r$. The sampling designs of these samples are independent and

$$P_{1a}(s_1) = \frac{1}{\binom{i-1}{r-1}}, \quad P_{1b}(s_2) = \frac{1}{\binom{N-i}{n-r}} \tag{3.23}$$

Hence, the selected sample is $s = \{s_1, i, s_2\}$ and its probability is

$$P_{*,r}(i|u, v)P_{1a}(s_1)P_{1b}(s_2) = P_r(s|u, v)$$

where $r = u, u + 1, ..., v$.

The ratio quantile-type estimator is defined as follows:

$$\bar{y}_{rqs} = \bar{y}_s \frac{E_0(X_{(r)})}{X_{(r)}},$$ (3.24)

where $E_0(X_{(r)})$ is given by the expression (3.3).

Theorem 3.3 *If $u = r$ and $v = N - n + r$, the strategy $(t_{rs}, P_r(s))$ leads to an unbiased estimation of the population mean.*

3.3 Sampling Design Proportional to Function of Two-Order Statistics

Let $x_i \leq x_j$ for $i < j$ and $i, j = 1, ..., N$ as it was previously assumed. Moreover, let $f(x_j, x_j)$ be a positive function of values x_i and x_j of the order statistics $X_{(r_1)}$ and $X_{(r_2)}$, respectively, where

$$r_1 \leq i \leq N - n + r_1 \quad and \quad r_1 < r_2 \leq j \leq N - n + r_2.$$

Moreover, let

$$z(r_1, r_2) = \sum_{i=r_1}^{N-n+r_1} \sum_{j=i+r_2-r_1}^{N-n+r_2} f(x_i, x_j)g(r_1, r_2, i, j).$$

Wywiał (2009a, b, 2011) analyzed the sampling design proportional to the following difference of order statistics:

$$f(x_i, x_j) = x_j - x_i$$ (3.25)

where x_i and x_j are values of the order statistics $X_{(r_1)}$ and $X_{(r_2)}$, respectively. In particular, when $r_1 = 1$ and $r_2 = n$, the sampling design is proportional to the sample range of an auxiliary variable. The straightforward generalization of the sampling design proportional to $x_j - x_i$ is as follows.

Definition 3.3 The sampling design proportional to the non-negative function $f(x_i, x_j)$ of the order statistics $X_{(r_2)}, X_{(r_1)}$ is as follows:

$$P_{r_1, r_2}(s) = \frac{f(x_i, x_j)}{z(r_1, r_2)} \quad (i, j) \in s \in G(r_1, r_2, i, j).$$ (3.26)

To define the conditional sampling design let us introduce the following:

$$f(x_j, x_i, C) = \begin{cases} f(x_i, x_j) & for \quad f(x_i, x_j) \in C, \\ 0 & for \quad f(x_i, x_j) \notin C \end{cases} \tag{3.27}$$

where C is a non-empty set of the positive real numbers, so $C \subseteq R_+$. Moreover, let

$$z(r_1, r_2, C) = \sum_{i=r_1}^{N-n+r_1} \sum_{j=i+r_2-r_1}^{N-n+r_2} f(x_i, x_j, C) g(r_1, r_2, i, j). \tag{3.28}$$

Definition 3.4 The conditional sampling design proportional to the non-negative function $f(x_i, x_j, c)$ of the order statistics $X_{(r_2)}, X_{(r_1)}$ is as follows:

$$P_{r_1, r_2}(s|C) = \frac{f(x_i, x_j, C)}{z(r_1, r_2, C)} \qquad (i, j) \in s \in G(r_1, r_2, i, j). \tag{3.29}$$

Both the above definitions lead to the conclusion that for $C = R_+$ the conditional sampling design $P_{r_1, r_2}(s|C)$ reduces to the unconditional one $P_{r_1, r_2}(s)$.

Theorem 3.4 *Under the sampling design $P_{r_1, r_2}(s|C)$, the inclusion probabilities of the first order are as follows:*

$$\pi_k(r_1, r_2, C) =$$

$$\frac{1}{z(r_1, r_2, C)} \Bigg(\delta(r_1 - 1) \sum_{i=r_1\delta(r_1-k)+(k-1)\delta(k+1-r_1)\delta(N-n+r_1-k)}^{N-n+r_1} \sum_{j=i+r_2-r_1}^{N-n+r_2} \binom{i-2}{r_1-2}$$

$$\binom{j-i-1}{r_2-r_1-1}\binom{N-j}{n-r_2} f(x_j, x_i, C) + \delta(k-r_1)\delta(N-n+r_2-k)\delta(r_2-r_1-1)$$

$$\sum_{i=r_1}^{min(k-1,N-n+r_1)} \sum_{j=max(i+r_2-r_1,k+1)}^{N-n+r_2} \binom{i-1}{r_1-1}\binom{j-i-2}{r_2-r_1-2}\binom{N-j}{n-r_2} f(x_j, x_i, C)+$$

$$\delta(k-r_2)\delta(n-r_2)\delta(N-n+r_2-k+1)$$

$$\sum_{i=r_1}^{k-r_2+r_1-1} \sum_{j=i+r_2-r_1}^{k-1} \binom{i-1}{r_1-1}\binom{j-i-1}{r_2-r_1-1}\binom{N-j-1}{n-r_2-1} f(x_j, x_i, C) + \delta(n-r_2)$$

$$\delta(k-N+n-r_2) \sum_{i=r_1}^{N-n+r_1} \sum_{j=i+r_2-r_1}^{N-n+r_2} \binom{i-1}{r_1-1}\binom{j-i-1}{r_2-r_1-1}\binom{N-j-1}{n-r_2-1} f(x_j, x_i, C)+$$

$$\delta(k+1-r_1)\delta(N-n+r_1-k+1)\binom{k-1}{r_1-1}$$

$$\sum_{j=k+r_2-r_1}^{N-n+r_2}\binom{j-k-1}{r_2-r_1-1}\binom{N-j}{n-r_2}f(x_j,x_k,C)+\delta(k-r_2+1)\delta(N-n+r_2-k+1)$$

$$\binom{N-k}{n-r_2}\sum_{i=r_1}^{k-r_2+r_1}\binom{i-1}{r_1-1}\binom{k-i-1}{r_2-r_1-1}f(x_k,x_i,C)\right) \tag{3.30}$$

Corollary 3.1 *Under the sampling design* $P_{1,n}(s|C)$, *the inclusion probabilities of the first-order are as follows:*

$$\pi_k(1,n,C)=\frac{1}{z(1,n,C)}\left(\delta(k-1)\delta(N-k)\delta(n-2)\sum_{i=1}^{min(k-1,N-n+1)}\sum_{j=max(i+n-1,k+1)}^{N}\right.$$

$$\binom{j-i-2}{n-3}f(x_j,x_i,C)+\delta(N-n+2-k)\sum_{j=k+n-1}^{N}\binom{j-k-1}{n-2}f(x_j,x_k,C)+$$

$$\delta(k-n+1)\delta(N+1-k)\sum_{i=1}^{k-n+1}\binom{k-i-1}{n-2}f(x_k,x_i,C)\right) \tag{3.31}$$

Theorem 3.5 *The inclusion second-order probabilities of the sampling design* $P_{r_1,r_2}(s|C)$ *are as follows:*

$$\pi_{k,t}(r_1,r_2,C)=P(k,t\in s_1)+P\left(k\in s_1,X_{(r_1)}=x_t\right)+P(k\in s_1,t\in s_2)+$$
$$P\left(k\in s_1,X_{(r_2)}=x_t\right)+P(k\in s_1,t\in s_3)+P\left(X_{(r_1)}=x_k,t\in s_2\right)+$$
$$P\left(X_{(r_1)}=x_k,X_{(r_2)}=x_t\right)+P\left(X_{(r_1)}=x_k,t\in s_3\right)+P(k,t\in s_2)+$$
$$P\left(k\in s_2,X_{(r_2)}=x_t\right)+P(k\in s_2,t\in s_3)+P\left(X_{(r_2)}=x_k,t\in s_3\right)+P(k,t\in s_3) \tag{3.32}$$

where

$$P(k,t\in s_1)=\frac{\delta(r_1-2)\delta(N-n+r_1-t)}{z(r_1,r_2)}.$$

$$\sum_{i=max(r_1,t+1)}^{N-n+r_1}\sum_{j=i+r_2-r_1}^{N-n+r_2}\binom{i-3}{r_1-3}\binom{j-i-1}{r_2-r_1-1}\binom{N-j}{n-r_2}f(x_j,x_i,C).$$

$$\tag{3.33}$$

$$P\left(k \in s_1, X_{(r_1)} = x_t\right) =$$
$$\frac{\delta(r_1 - 1)\delta(N - n + r_1 - k)\delta(N - n + r_1 + 1 - t)\delta(t + 1 - r_1)}{z(r_1, r_2)} \cdot$$
$$\binom{t - 2}{r_1 - 2} \sum_{j=t+r_2-r_1}^{N-n+r_2} \binom{j - t - 1}{r_2 - r_1 - 1}\binom{N - j}{n - r_2} f(x_j, x_t, C), \quad (3.34)$$

$$P\left(k \in s_1, t \in s_2\right) =$$
$$\frac{\delta(N - n + r_1 - k)\delta(t - r_1)\delta(r_1 - 1)\delta(r_2 - r_1 - 1)\delta(N - n + r_2 - t)\delta(t - k - 1)}{z(r_1, r_2)} \cdot$$
$$\sum_{i=max(r_1,k+1)}^{min(t-1,N-n+r_1)} \sum_{j=max(t+1,i+r_2-r_1)}^{N-n+r_2} \binom{i - 2}{r_1 - 2}\binom{j - i - 2}{r_2 - r_1 - 2}\binom{N - j}{n - r_2} f(x_j, x_i, C),$$
$$(3.35)$$

$$P\left(k \in s_1, X_{(r_2)} = x_t\right) =$$
$$\frac{\delta(N - n + r_1 - k)\delta(t + 1 - r_2)\delta(N - n + r_2 + 1 - t)\delta(t - k - r_2 + r_1)\delta(r_1 - 1)}{z(r_1, r_2)} \cdot$$
$$\binom{N - t}{n - r_2} \sum_{i=max(r_1,k+1)}^{t-r_2+r_1} \binom{i - 2}{r_1 - 2}\binom{t - i - 1}{r_2 - r_1 - 1} f(x_t, x_i, C), \quad (3.36)$$

$$P\left(k \in s_1, t \in s_3\right) =$$
$$\frac{\delta(N - n + r_1 - k)\delta(t - r_2)\delta(r_1 - 1)\delta(n - r_2)\delta(t - k - r_2 + r_1 - 1)}{z(r_1, r_2)} \cdot$$
$$\sum_{i=max(r_1,k+1)}^{min(t-r_2+r_1-1,N-n+r_1)} \sum_{j=i+r_2-r_1}^{min(t-1,N-n+r_2)} \binom{i - 2}{r_1 - 2}\binom{j - i - 1}{r_2 - r_1 - 1}\binom{N - j - 1}{n - r_2 - 1} f(x_j, x_i, C),$$
$$(3.37)$$

$$P\left(X_{(r_1)} = x_k, t \in s_2\right) =$$
$$\frac{\delta(k + 1 - r_1)\delta(N - n + r_1 + 1 - k)\delta(t - r_1)\delta(N - n + r_2 - t)\delta(r_2 - r_1 - 1)}{z(r_1, r_2)} \cdot$$
$$\binom{k - 1}{r_1 - 1} \sum_{j=max(t+1,k+r_2-r_1)}^{N-n+r_2} \binom{j - k - 2}{r_2 - r_1 - 2}\binom{N - j}{n - r_2} f(x_j, x_k, C), \quad (3.38)$$

$$P\left(X_{(r_1)} = x_k, X_{(r_2)} = x_t\right) =$$
$$\frac{\delta(k+1-r_1)\delta(N-n+r_1+1-k)\delta(t+1-r_2)\delta(N-n+r_2-t+1)}{z(r_1,r_2)}\cdot$$
$$\delta(t+1-k-r_2+r_1)\binom{k-1}{r_1-1}\binom{t-k-1}{r_2-r_1-1}\binom{N-t}{n-r_2}f(x_t,x_k,C), \quad (3.39)$$

$$P\left(X_{(r_1)} = x_k, t \in s_3\right) =$$
$$\frac{\delta(k+1-n)\delta(N-n+r_1+1-k)\delta(t-r_2)\delta(n-r_2)\delta(t-k-r_2+r_1)}{z(r_1,r_2)}\cdot$$
$$\binom{k-1}{r_1-1}\sum_{j=k+r_2-r_1}^{min(t-1,N-n+r_2)}\binom{j-k-1}{r_2-r_1-1}\binom{N-j-1}{n-r_2-1}f(x_j,x_k,C), \quad (3.40)$$

$$P\left(k, t \in s_2\right) = \frac{\delta(k-r_1)\delta(N-n+r_2-t)\delta(r_2-r_1-2)}{z(r_1,r_2)}\cdot$$
$$\sum_{i=r_1}^{min(k-1,N-n+r_1)}\sum_{j=max(t+1,i+r_2-r_1)}^{N-n+r_2}\binom{i-1}{r_1-1}\binom{j-i-3}{r_2-r_1-3}\binom{N-j}{n-r_2}f(x_j,x_i,C),$$
$$(3.41)$$

$$P\left(k \in s_2, X_{(r_2)} = x_t\right) =$$
$$\frac{\delta(k-r_1)\delta(N-n+r_2-k)\delta(t+1-r_2)\delta(N-n+r_2+1-t)\delta(r_2-r_1-1)}{z(r_1,r_2)}\cdot$$
$$\binom{N-t}{n-r_2}\sum_{i=r_1}^{min(k-1,t-r_2+r_1)}\binom{i-1}{r_1-1}\binom{t-i-2}{r_2-r_1-2}f(x_t,x_i,C), \quad (3.42)$$

$$P\left(k \in s_2, t \in s_3\right) =$$
$$\frac{\delta(k-r_1)\delta(N-n+r_2-k)\delta(t-r_2)\delta(t-k-1)\delta(r_2-r_1-1)\delta(n-r_2)}{z(r_1,r_2)}\cdot$$
$$\sum_{i=r_1}^{min(k-1,t-r_2+r_1-1,N-n+r_1)}\sum_{j=max(i+r_2-r_1,k+1)}^{min(t-1,N-n+r_2)}\binom{i-1}{r_1-1}\binom{j-i-2}{r_2-r_1-2}\binom{N-j-1}{n-r_2-1}\cdot$$
$$f(x_j,x_i,C), \quad (3.43)$$

$$P\left(X_{(r_2)} = x_k, t \in s_3\right) = \frac{\delta(k+1-r_2)\delta(N-n+r_2+1-k)\delta(t-r_2)\delta(n-r_2)}{z(r_1,r_2)} \cdot$$

$$\binom{N-k-1}{n-r_2-1} \sum_{i=r_1}^{k-n+r_1} \binom{i-1}{r_1-1}\binom{k-i-1}{r_2-r_1-1} f(x_k, x_i, C), \quad (3.44)$$

$$P(k, t \in s_3) = \frac{\delta(k-r_2)\delta(n-r_2-1)}{z(r_1,r_2)} \cdot$$

$$\sum_{i=r_1}^{min(N-n+r_1,k-1-r_2+r_1)} \sum_{j=i+r_2-r_1}^{min(k-1,N-n+r_2)} \binom{i-1}{r_1-1}\binom{j-i-1}{r_2-r_1-1}\binom{N-j-2}{n-r_2-2} \cdot$$

$$f(x_j, x_i, C). \quad (3.45)$$

The sampling scheme implementing the sampling design $P_{r_1,r_2}(s|C)$ is as follows. First, population elements are ordered according to increasing values of the auxiliary variable. Let

$$s = s_1 \cup \{i\} \cup s_2 \cup \{j\} \cup s_3, \qquad s_1 = \{k : k \in U, x_k < x_i\},$$

$$s_2 = \{k : k \in U, x_j > x_k > x_i\} \quad \text{and} \quad s_3 = \{k : k \in U, x_k > x_j\}$$

Moreover, let

$$U = U(1, i-1) \cup \{i\} \cup U(i+1, j-1) \cup \{j\} \cup U(j+1, N)$$

where

$$U(1, i-1) = (1, ..., i-1), \qquad U(i+1, j-1) = (i+1, ..., j-1),$$

$$U(j+1, N) = (j+1, ..., N).$$

Let $\mathbf{S}_1 = \mathbf{S}(U(1, i-1); s_1)$ be sample space of the sample s_1 of size $r_1 - 1$, $\mathbf{S}_2 = \mathbf{S}(U(i+1, j-1); s_2)$ be sample space of the sample s_2 of size $r_2 - r_1 - 1$, $\mathbf{S}_3 = \mathbf{S}(U(j+1, N); s_3)$ be sample space of the sample s_3 of size $n - r_2$. Similarly, $\mathbf{S} = \mathbf{S}(U, s)$).

The sampling scheme is given by the following probabilities:

$$P_{r_1,r_2}(s|C) = P_{1a}(s_1) p_{r_1,r_2}(i|C) P_{1b}(s_2) p'_{r_1,r_2}(j|C) P_{1c}(s_3) \quad (3.46)$$

where

$$P_{1a}(s_1) = \binom{i-1}{r_1-1}^{-1}, \quad P_{1b}(s_2) = \binom{j-i-1}{r_2-r_1-1}^{-1}, \quad P_{1c}(s_3) = \binom{N-j}{n-r_2}^{-1}, \tag{3.47}$$

$$s_1 \in \mathbf{S}_1, \qquad s_2 \in \mathbf{S}_2, \qquad s_3 \in \mathbf{S}_3,$$

$$p_{r_1,r_2}(i|j,C) = P\left(X_{(r_1)} = x_i | X_{(u)} = x_j, C\right) = \frac{P_{r_1,u}\left(X_{(r_1)} = x_i, X_{(u)} = x_j, C\right)}{P_{r_1,u}\left(X_{(u)} = x_j, C\right)}, \tag{3.48}$$

$$P_{r_1,r_2}\left(X_{(r_1)} = x_i, X_{(r_2)} = x_j, C\right) = \sum_{s \in G(r_1,r_2,i,j)} P_{r_1,r_2}(s|C) = \frac{f(x_j, x_i, C)g(r_1, r_2, i, j)}{z(r_1, r_2, C)}, \tag{3.49}$$

$$p'_{r_1,r_2}(j|C) = P_{r_1,r_2}\left(X_{(r_2)} = x_j, C\right) = \frac{1}{z(r_1, r_2, C)} \sum_{i=r_1}^{N-n+r_1} f(x_j, x_i, C)g(r_1, r_2, i, j). \tag{3.50}$$

To select the sample s, first, the jth element of the population is selected according to the probability function $p'_{r_1,r_2}(j|C)$. Next, the ith element of the population is drawn according to the probability function $p_{r_1,r_2}(i|j, C)$. Finally, the samples s_1, s_2, and s_3 are selected according to the sampling designs $P_{1a}(s_1)$, $P_{1b}(s_2)$, and $P_{1c}(s_3)$, respectively.

Let us assume that a variable under study is explained by an auxiliary variable by means of the following equation $y_i = a + bx_i + e_i$ for all $i \in U$, and $\sum_{i \in U} e_i = 0$. The residuals of that linear regression function are not correlated with the auxiliary variable. Let $\left(X_{(r_1)}, Y_{[r_1]}\right)$ be a two-dimensional random variable where $X_{(r_1)}$ is the r_1th-order statistic of the auxiliary variable and $Y_{[r_1]}$ is concomitant of $X_{(r_1)}$, see David and Nagaraja (2003), p. 144. It allows us consider the following regression-type estimator:

$$\bar{y}_{regHTqs} = \bar{y}_{HTs} + b_{r_1,r_2,s} \left(\bar{x} - \bar{x}_{HTs}\right) \tag{3.51}$$

where

$$b_{r_1,r_2,s} = \frac{Y_{[r_2]} - Y_{[r_1]}}{X_{(r_2)} - X_{(r_1)}}. \tag{3.52}$$

Particularly, in the case of the simple random sample Design, we have the following:

$$\bar{y}_{regqs} = \bar{y}_s + \frac{Y_{[r_2]} - Y_{[r_1]}}{X_{(r_2)} - X_{(r_1)}} \left(\bar{x} - \bar{x}_s\right).$$

Let

$$b_{r_1,r_2} = \frac{E_x(Y_{[r_2]}) - E_x(Y_{[r_1]})}{E(X_{(r_2)}) - E(X_{(r_1)})}$$

where $E_x(Y_{[r_1]}) = \sum_{k=1}^{N} y_k P(X_{(r_1)} = x_k)$, $E_x(Y_{[r_2]}) = \sum_{k=1}^{N} y_k P(X_{(r_2)} = x_k)$ and $E(X_{(r_2)})$, $E(X_{(r_1)})$ are explained by the expression (3.3).

Theorem 3.6 *Under the sampling design stated in the Definition 3.4, the parameters of the strategy* $\left(\bar{y}_{regHTqs}, P_{r_2,r_1}^{(-)}(s) \right)$ *are approximately as follows:*

$$E\left(\bar{y}_{regHTqs}, P_{r_2,r_1}^{(-)}(s) \right) \approx \bar{y},$$

$$V\left(\bar{y}_{regHTqs}, P_{r_2,r_1}^{(-)}(s) \right) \approx V\left(\bar{y}_{HTs}, P_{r_2,r_1}^{(-)}(s) \right) - 2b_{r_1,r_2} Cov\left(\bar{y}_{HTs}, \bar{x}_{HTs}, P_{r_2,r_1}^{(-)}(s) \right) +$$
$$b_{r_1,r_2}^2 V\left(\bar{x}_{HTs}, P_{r_2,r_1}^{(-)}(s) \right) \qquad (3.53)$$

where

$$Cov\left(\bar{y}_{HTs}, \bar{x}_{HTs}, P_{r_2,r_1}^{(-)}(s) \right) = \frac{1}{N^2} \left(\sum_{k \in U} \sum_{l \in U} \Delta_{k,l} \frac{y_k}{\pi_k} \frac{x_l}{\pi_l} \right), \qquad (3.54)$$

$$\Delta_{k,l} = \pi_{k,l} - \pi_k \pi_l, \quad V\left(\bar{x}_{HTs}, P_{r_2,r_1}^{(-)}(s) \right) = Cov\left(\bar{x}_{HTs}, \bar{x}_{HTs}, P_{r_2,r_1}^{(-)}(s) \right),$$
$$V\left(\bar{y}_{HTs}, P_{r_2,r_1}^{(-)}(s) \right) = Cov\left(\bar{y}_{HTs}, \bar{y}_{HTs}, P_{r_2,r_1}^{(-)}(s) \right).$$

The estimator of the variance: $V\left(\bar{y}_{regHTqs}, P_{r_2,r_1}^{(-)}(s) \right)$ is as follows:

$$V_s\left(\bar{y}_{regHTqs}, P_{r_2,r_1}^{(-)}(s) \right) = V_s\left(\bar{y}_{HTs}, P_{r_2,r_1}^{(-)}(s) \right) - 2b_{r_1,r_2,s} Cov_s\left(\bar{y}_{HTs}, \bar{x}_{HTs}, P_{r_2,r_1}^{(-)}(s) \right) +$$
$$+ b_{r_1,r_2,s}^2 V_s\left(\bar{x}_{HTs}, P_{r_2,r_1}^{(-)}(s) \right) \qquad (3.55)$$

where

$$Cov_s\left(\bar{y}_{HTs}, \bar{x}_{HTs}, P_{r_2,r_1}^{(-)}(s) \right) = \frac{1}{N^2} \left(\sum_{k \in s} \sum_{l \in s} \Delta_{*,k,l} \frac{y_k}{\pi_k} \frac{x_l}{\pi_l} \right), \qquad (3.56)$$

$$\Delta_{*k,l} = \frac{\Delta_{k,l}}{\pi_{k,l}}, \quad V_s\left(\bar{x}_{HTs}, P_{r_2,r_1}^{(-)}(s) \right) = Cov_s\left(\bar{x}_{HTs}, \bar{x}_{HTs}, P_{r_2,r_1}^{(-)}(s) \right),$$
$$V_s\left(\bar{y}_{HTs}, P_{r_2,r_1}^{(-)}(s) \right) = Cov_s\left(\bar{y}_{HTs}, \bar{y}_{HTs}, P_{r_2,r_1}^{(-)}(s) \right).$$

In the case of the ordinary regression strategy defined in the Sect. 1.4, the probability that the sample variances of the auxiliary variable is close to zero can be large. In this situation, the slope coefficient of the ordinary regression function can have large spread, and consequently, the regression statistic, defined by the expressions (1.14), can be an inaccurate estimator of the population mean.

In this situation, Wywiał (2003) considered regression estimator of a population mean under the conditional version of Singh Srivastava's (1980) sampling design. The sampling design is proportional to the sample variance v_s under the condition that $v_s > c$. In this situation, the appropriate simulation analysis let us infer that the accuracy of the regression estimator usually increases when the value c increases. But the sampling strategy is not convenient in practical research because derivations of the inclusion probabilities are rather complicated. In this situation, the following conditional sampling design can be considered.

$$P_{n,1}^{(-)}\left(s|X_{(n)} - X_{(1)} > c\right) = \frac{x_j - x_i}{z\left(1, n|X_{(n)} - X_{(1)} > c\right)}, \qquad (i, j) \in s \in G(1, n, i, j)$$

(3.57)

where $z\left(1, n|X_{(n)} - X_{(1)} > c\right)$ is defined by the expression (3.28), x_j and x_i are values of the order statistics $X_{(n)}$ and $X_{(1)}$, respectively, and $x_j - x_i > c$, $c \geq 0$. Under this sampling design, the denominator of the estimator of the regression slope coefficient should not be close to zero. Hence, in consequence, this can improve the accuracy of the regression estimator. In Chap. 4, the efficiency of the conditional regression strategy will be considered.

Wywiał (2016) prepared programmes written in the R, which facilitated evaluation of the inclusion probabilities of the first and second orders provided by the theorems 3.1–3.5.

3.4 Sampling Design Proportional to Function of Three-Order Statistics

First, the properties of the sampling design dependent on the function of the three-order statistics of an auxiliary variable from the simple random sample drawn without replacement of size n will be considered.

Similarly, as in the previous sections, let $f(x_{i_1}, x_{i_2}, x_{i_3})$ be a non-negative and increasing function of values $x_{i_1}, x_{i_2}, x_{i_3}$ of the order statistics $X_{(r_1)}, X_{(r_2)}, X_{(r_3)}$, respectively, and $r_j \leq N - n + r_j$, $j = 1, 2, 3$. Moreover, let

$$f(x_{i_1}, x_{i_2}, x_{i_3}, C) = \begin{cases} f(x_{i_1}, x_{i_2}, x_{i_3}) & for \quad f(x_{i_1}, x_{i_2}, x_{i_3}) \in C, \\ 0 & for \quad f(x_{i_1}, x_{i_2}, x_{i_3}) \notin C \end{cases}$$

(3.58)

where $C \subseteq R_+$,

$$z_{r_1,\dots,r_h,C} = \sum_{i_1=1}^{N-n+r_1} \sum_{i_2=i_1+r_2-r_1}^{N-n+r_2} \dots \sum_{i_3=i_2+r_3-r_2}^{N-n+r_3} g(r_1, r_2, r_3; i_1, i_2, i_3) f(x_{i_1}, x_{i_2}, x_{i_3}, C).$$

(3.59)

Definition 3.5 The sampling design proportional to the function $f(x_{i_1}, x_{i_2}, x_{i_3}, C)$ is as follows:

$$P_{r_1,r_2,r_3}(s|C) = \frac{f(x_{i_1}, x_{i_2}, x_{i_3}, C)}{z_{r_1,r_2,r_3,C}}$$

(3.60)

for $(i_1, i_2, i_3) \in s \in G(r_1, r_2, r_3; i_1, i_2, i_3)$.

The sampling design $P_{r_1,r_2,r_3}(s|C)$ reduces to the unconditional one, denoted by $P_{r_1,r_2,r_3}(s)$, when $C = R_+$.

Theorem 3.7 *Under the sampling design $P_{r_1,r_2,r_3}(s|C)$, the first-order inclusion probabilities are as follows.*

$$\pi_k(r_1, r_2, r_3, C)$$

$$= \frac{1}{z_{r_1,r_2,r_3,C}}\Bigg(\delta(r_1 - 1) \sum_{i_1=r_1\delta(r_1-k)+\delta(k-r_1+1)\delta(N-n+r_1-k)(k+1)}^{N-n+r_1} \sum_{i_2=i_1+r_2-r_1}^{N-n+r_2} \sum_{i_3=i_2+r_3-r_2}^{N-n+r_3}$$

$$\frac{\binom{i_1-2}{r_1-2}}{\binom{i_1-1}{r_1-1}} g(r_1, r_2, r_3; i_1, i_2, i_3) f(i_1, i_2, i_3, C)$$

$$+ \delta(k - r_1)\delta(r_2 - r_1 - 1)\delta(N - n + r_2 - k) \sum_{i_1=r_1}^{min(N-n+r_1,k-1)} \sum_{i_2=max(i_1+r_2-r_1,k+1)}^{N-n+r_2}$$

$$\sum_{i_3=i_2+r_3-r_2}^{N-n+r_3} \frac{\binom{i_2-i_1-2}{r_2-r_1-2}}{\binom{i_2-i_1-1}{r_2-r_1-1}} g(r_1, r_2, r_3; i_1, i_2, i_3) f(i_1, i_2, i_3, C)$$

$$+ \delta(k - r_2)\delta(N - n + r_3 - k)\delta(r_3 - r_2 - 1) \sum_{i_1=r_1}^{min(N-n+r_1,k+r_1-r_2-1)} \sum_{i_2=i_1+r_2-r_1}^{min(N-n+r_2,k-1)}$$

$$\sum_{i_3=max(i_2+r_3-r_2,k+1)}^{N-n+r_3} \frac{\binom{i_3-i_2-2}{r_3-r_2-2}}{\binom{i_3-i_2-1}{r_3-r_2-1}} g(r_1, r_2, r_3; i_1, i_2, i_3) f(i_1, i_2, i_3, C)$$

$$+ \delta(n - r_3)\delta(k - r_3)\delta(N - n + r_3 + 1 - k) \sum_{i_1=r_1}^{k+r_1-r_3-1} \sum_{i_2=i_1+r_2-r_1}^{k+r_2-r_3-1} \sum_{i_3=i_2+r_3-r_2}^{k-1} \frac{\binom{N-i_3-1}{n-r_3-1}}{\binom{N-i_3}{n-r_3}}$$

$$g(r_1, r_2, r_3; i_1, i_2, , i_3) f(i_1, i_2, i_3, C)$$

$$+ \delta(n - r_3 - 1)\delta(k - N + n - r_3) \sum_{i_1=r_1}^{N-n+r_1} \sum_{i_2=i_1+r_2-r_1}^{N-n+r_2} \sum_{i_3=i_2+r_3-r_2}^{N-n+r_3} \frac{\binom{N-i_3-1}{n-r_3-1}}{\binom{N-i_3}{n-r_3}}$$

$$g(r_1, r_2, r_3; i_1, i_2, i_3) f(i_1, i_2, i_3, C)$$

$$+ \delta(N - n + r_1 + 1 - k)\delta(k - r_1 + 1)\binom{k-1}{r_1-1} \sum_{i_2=k+r_2-r_1}^{N-n+r_2} \sum_{i_3=i_2+r_3-r_2}^{N-n+r_3} \binom{i_2-k-1}{r_2-r_1-1}$$
$$\binom{i_3-i_2-1}{r_3-r_2-1}\binom{N-i_3}{n-r_2} f(k, i_2, i_3, C)$$

$$+ \delta(N - n + r_2 + 1 - k)\delta(k - r_2 + 1) \sum_{i_1=r_1}^{k-r_2+r_1} \sum_{i_3=k+r_3-r_2}^{N-n+r_3} \binom{i_1-1}{r_1-1}\binom{k-i_1-1}{r_2-r_1-1}$$
$$\binom{i_3-k-1}{r_3-r_2-1}\binom{N-i_3}{n-r_3} f(i_1, k, i_3, C)$$

$$+\delta(N - n + r_3 + 1 - k)\delta(k - r_3 + 1) \sum_{i_1=r_1}^{k+r_1-r_3} \sum_{i_2=i_1+r_2-r_1}^{k+r_2-r_3}$$
$$\binom{i_1-1}{r_1-1}\binom{i_2-i_1-1}{r_2-r_1-1}\binom{k-i_2-1}{r_3-r_2-1}\binom{N-k}{n-r_3} f(i_1, i_2, k, C)\Bigg). \quad (3.61)$$

The theorem can be proved similar to Theorem 3.4.

Sampling Scheme

The construction of the sampling scheme implementing the sampling design proposed by the Definition 3.5 is as follows. First, the sampling design $P_{r_1,r_2,r_3}(s|C)$ can be rewritten in the following way:

$$P_{r_1,r_2,r_3}(s|C) = p(i_1, i_2, i_3|C)P(s_1)P(s_2)P(s_3)P(s_4) \quad (3.62)$$

where

$$p(i_1, i_2, i_3|C) = \frac{f(i_1, i_2, i_3, C)g(r_1, r_2, r_3; i_1, i_2, i_3)}{z_{r_1,r_2,r_3}} \quad (3.63)$$

$$P(s_1) = \frac{1}{\binom{i_1-1}{r_1-1}}, \quad , P(s_2) = \frac{1}{\binom{i_2-i_1-1}{r_2-r_1-1}} \quad P(s_3) = \frac{1}{\binom{i_3-i_2-1}{r_3-r_2-1}}, \quad P(s_4) = \frac{1}{\binom{N-i_3}{n-r_3}}$$
$$(3.64)$$

where

$$s_1 \in S(U(1, ..., i_1 - 1), s_1), \quad s_2 \in S(U(i_1 + 1, ..., i_2 - 1), s_2),$$

$$s_3 \in S(U(i_2 + 1, ..., i_3 - 1), s_3), \quad s_4 \in S(U(i_3 + 1, ..., N), s_1),$$

$$r_j \le i_j \le N - n + r_j, \text{ and } i_j < i_{j+1}, j \le 3.$$

Now the set of the values $(x_{i_1}, x_{i_2}, x_{i_3})$ is drawn with the probability $p(x_{i_1}, x_{i_2}, x_{i_3}|C)$ where

$$p(i_1, i_2, i_3|C) = p(i_1|i_2, i_3, C)p(i_2|i_3, C)p(i_3|C) \qquad (3.65)$$

$$p(i_1|i_2, i_3, C) = \frac{p(i_1, i_2, i_3, C)}{p(i_2, i_3|C)}, \quad i_2 > i_1, \quad p(i_2|i_3, C) = \frac{p(i_2, i_3|C)}{p(i_3|C)}, \quad i_3 > i_2, \qquad (3.66)$$

$$p(i_2, i_3|C) = \sum_{i_1=r_1}^{N-n+r_1} p(i_1, i_2, i_3|C), \quad p(i_3|C) = \sum_{i_1=r_1}^{N-n+r_1} \sum_{i_2=i_1+r_2-r_1}^{N-n+r_2} p(i_1, i_2, i_3|C).$$

$$(3.67)$$

Hence, the element i_3 is drawn without replacement from the population with the probability $p(i_3|C)$. Next, the element i_2 is selected without replacement from the set $U - \{i_3\}$ with the conditional probability $p(i_2|i_3, C)$ and, finally, the element i_1 is drawn without replacement from the set $U - \{i_2, i_3\}$ with the conditional probability $p(i_1|i_2, i_3, C)$.

In the next step, the sequence of the simple random samples (s_1, s_2, s_3, s_4) are drawn in the following manner. The sample s_j is the simple random sample of the size $r_j - r_{j-1}$ drawn without replacement from the sup-population $U(i_{j-1}, ..., i_j)$ where $j = 1, ..., 4$ and $r_0 = 0$, $i_0 = 0$, $r_4 = n$, $i_4 = N$. The algorithm leads to the selection of the sample s

$$s = s_1 \cup \{i_1\} \cup s_2 \cup \{i_2\} \cup s_3 \cup \{i_3\} \cup s_4.$$

Considered sampling designs proportional to functions of order statistics can be developed in several directions. The introduced sampling designs should be generalized into the case of function of more than three-order statistics. But in this case, the expressions for the inclusion probabilities became very complicated. In the next chapter, many application of the introduced sampling designs are presented.

References

David, H. A., & Nagaraja, H. N. (2003). Order Statistics. Hoboken: Wiley

Guenther, W. (1975). The inverse hypergeometric—a useful model. *Statistica Neerlandica, 29,* 129–144.

Singh, P., Srivastava, & A. K. (1980). Sampling schemes providing unbiased regression estimators. *Biometrika, 67*(1), 205–209.

Tillé, Y.(1998). Estimation in surveys using conditional inclusion probabilities: simple random sampling. *International Statistical Review, 66*(3), 303–322.

Tillé, Y. (2006). *Sampling Algorithms*. Berlin: Springer.

Wilks, S. S. (1962). *Mathematical Statistics*. New York, London: Wiley.

WywiałJ, L. (2003). On conditional sampling strategies. *Statistical Papers, 44*(3), 397–419.

Wywiał, J. L. (2008). Sampling design proportional to order statistic of auxiliary variable. *Statistical Papers, 49*(2), 277–289.

Wywiał, J. L. (2009a). Performing quantiles in regression sampling strategy. *Model Assisted Statistics and Applications, 4*(2), 131–142.

Wywiał, J. (2009b). Sampling design proportional to positive function of order statistics of auxiliary variable. *Studia Ekonomiczne-Zeszyty Naukowe, 53*, 35–60.

WywiałJ, L. (2011). Sampling designs proportionate to sum of two order statistics of auxiliary variables. *Statistics in Transition—New Series, 14*(2), 231–248.

Wywiał, J. L. (2016). *Contributions to Testing Statistical Hypotheses in Auditing*. Warsaw: PWN.

Chapter 4
Simulation Analysis of the Efficiency of the Strategies

This chapter compares the accuracy of estimation of the population mean using Monte Carlo simulation studies. Several sampling strategies are considered which are treated as pairs consisting of estimators and sampling plans. In particular, the accuracy of conditional sampling designs is studied. Problems of estimation of domain means are treated separately. Moreover, the accuracy of quantile estimators from sampling designs dependent on order statistics is also considered.

4.1 Description of the Simulation Experiments

The comparison of the estimation strategies is based on their mean square errors or variances. To calculate those parameters, we assume that all values of the variable under study as well as the values of an auxiliary population are known in the whole population. We consider an artificial population where the observed variables are generated as values of random variables whose probability distribution are defined in advance. The two-dimensional normal random variable as well as the two-dimensional log-normal random variable are taken into account. The former has a symmetrical probability distribution and the latter has a right-hand asymmetrical (skewed) probability distribution. Non-homogeneous data are also considered. In this case, the two-dimensional random variable is defined as a mixture of three normal distributions.

We consider the set of values (x_k, y_k), $k = 1, ..., N$, of the two-dimensional random variable (X, Y) which has a two-dimensional normal distribution denoted by $N(E(X), E(Y), V(X), V(Y), \rho)$ where ρ is the correlation coefficient. Moreover, we take into account the random variable (X, Y) which has a log-normal distribution such that $(\ln(X), \ln(Y)) \sim N(0, 0, 1, 1, \rho)$. Let us underline that we assume, as in the

© The Author(s), under exclusive license to Springer-Verlag GmbH, DE, part of Springer Nature 2021
J. L. Wywiał, *Sampling Designs Dependent on Sample Parameters of Auxiliary Variables*, SpringerBriefs in Statistics, https://doi.org/10.1007/978-3-662-63413-4_4

previous chapters, that observations of the auxiliary variable are ordered from the smallest to the largest, so $x_k \leq x_{k+1}$ for $k = 1, ..., N - 1$.

Let t_s be an estimator of the population mean $\bar{y}$ where the sample s is drawn according a sampling design $P(s)$. The mean square error of the estimator t_s is assessed in the following way. Let s_k, $k = 1, ..., K$, be the kth sample of a fixed size n sample drawn according to the sampling design $P(s)$ from a population of size $N > n$. The samples $\{s_k\}, k = 1, ..., K$, are drawn independently from the population. The mean square error of the statistic t_s is estimated by means of the following expression:

$$MSE(t_s) = \frac{1}{K} \sum_{k=1}^{K} (t_{s_k} - \bar{y})^2.$$

The accuracy of the estimation strategies was compared on the basis of the following coefficient of the relative efficiency:

$$deff(t_s, P(s)) = \frac{MSE(t_s, P(s))}{V(\bar{y}_s, P_0(s))} 100\% \tag{4.1}$$

where $V(\bar{y}_s, P_0(s))$ is the variance of the mean from the simple random sample drawn without replacement, given by expressions (1.5), (1.3).

Let us note that in the tables below the values of the efficiency coefficient are rounded to nearest integer number, so e.g. $deff(t_s, P(s)) = 0\%$ actually means $deff(t_s, P(s)) < 0.5\%$.

When $deff(t_s, P(s)) \leq 100\%$ ($deff(t_s, P(s)) < 100\%$), then we say that the estimation strategy $(t_s, P(s))$ is not less (is more) efficient than the simple random sample mean denoted by $(\bar{y}_s, P_0(s))$ or $(t_s, P(s))$ is not less (is more) accurate then $(\bar{y}_s, P_0(s))$. When $deff(t_{a,s}, P_a(s)) \leq deff(t_{b,s}, P_c(s))$, we say that the estimation strategy $(t_{a,s}, P_a(s))$ is not less efficient than $(t_{b,s}, P_c(s))$ or the estimation strategy $(t_{a,s}, P_a(s))$ is not less accurate than $(t_{b,s}, P_c(s))$ in the sense of the mean square error. If $deff(t_{a,s}, P_a(s)) < deff(t_{b,s}, P_c(s))$, then the estimation strategy $(t_{a,s}, P_a(s))$ is more efficient than $(t_{b,s}, P_c(s))$ or the estimation strategy $(t_{a,s}, P_a(s))$ is more accurate than $(t_{b,s}, P_c(s))$.

The strategies considered in the previous chapters are listed in the Table 4.1. Let us note that the conditional version of a strategy $(t_s, P(s))$ will be denoted by $(t_s, P(s|C))$ where t_s is an estimator and $P(s|C)$ is the sampling design under the assumption that a function of the order statistics of an auxiliary variable fulfils some property denoted by C. For instance, $C : X_{(1)} + X_{(n)} \geq \bar{x}$.

During the simulation analysis, the contribution of the squared bias of the strategies in terms of their mean square error was calculated according to the expression:

$$rb(t_s, P(s)) = \frac{(E(t_s, P(s)) - \bar{y})^2}{MSE(t_s, P(s))} 100\%$$

Table 4.1 The considered strategies

Strategy	Comments
$(\bar{y}_s, P_0(s))$	The ordinary simple random sample mean, see the expressions (1.3), (1.5)
$(\bar{y}_{HTs}, P_{Sd}(s))$	Horvitz–Thompson strategy under Sampford's sampling design, see the Sect. 2.2
$(\bar{y}_{HTs}, P_r(s))$	Horvitz–Thompson strategy under a sampling design proportional to the rth-order statistic, see the expr. (1.19), (3.11)
$\left(\bar{y}_{HTs}, P_{u,r}^{(-)}(s)\right)$	Horvitz–Thompson strategy under the sampling design proportional to the difference of uth- and rth-order statistics, $u > r$, see the expr. (1.19), (3.57) and Definition 3.3
$\left(\bar{y}_{HTs}, P_{r,u}^{(+)}(s)\right)$	Horvitz–Thompson strategy under the sampling design proportional to the sum of rth- and uth-order statistics, $u > r$, see the expr. (1.19) and Definition 3.3
$(\bar{y}_{rs}, P_0(s))$	The ordinary ratio estimator under the simple random sample, (1.3), (1.9)
$(\bar{y}_{rs}, P_{LMS}(s))$	The ordinary ratio estimator under the Lahiri–Midzuno–Sen's sampling design, (1.9), (2.1)
$(\bar{y}_{rHTs}, P(s))$	The ratio-Horvitz–Thompson estimation strategy, (1.24)
$(\bar{y}_{rqs}, P(s))$	The ratio quantile-type estimation strategy, (3.24)
$(\bar{y}_{regs}, P_0(s))$	The ordinary regression estimator under the simple random sample, (1.3), (1.14)
$(\bar{y}_{regs}, P_{SS}(s))$	The ordinary regression estimator under Singh–Srivastava's sampling design, (2.5), (1.14)
$(\bar{y}_{regHTs}, P(s))$	The regression Horvitz–Thompson estimation strategy, (1.28)
$(\bar{y}_{regHTqs}, P(s))$	The regression Horvitz–Thompson quantile estimation strategy, (3.51)

The values of the above coefficient are not presented below in the tables with results of the simulation analysis when they take values not greater than 1%.

The simulation analysis was supported by programs written by author, in the R. Several sub-programs that implemented the inclusion probabilities of the sampling designs, dependent on the order statistics, were published by Wywiał (2016).

4.2   Efficiency of Estimation Strategies Dependent on Sample Moments or Order Statistics

A simulation analysis of the accuracy of population mean estimation based on sampling design proportional to the value of the order statistic of the auxiliary variable is performed by Wywiał (2007). He considered the empirical data from Sweden's $N = 284$ municipalities that can be found in the monograph by Särndal et al. (1992). The variable under study y, was 1985 municipal taxation revenues, and the auxiliary variable x, was 1975 municipal population. The variables were right-skewed. The correlation coefficient between the variable under study and the auxiliary variable was equal to $\rho = 0.99$. In the population, there were three variable outlier observations. When the outliers are removed from the data, the correlation coefficient between the variable under study and the auxiliary variable is equal to $\rho = 0.97$.

The five strategies for estimation of the mean are taken into account: the simple random sample mean $(\bar{y}_s, P_0(s))$, the Horvitz–Thompson (1952) strategy $\left(\bar{y}_{HTs}, P_r(s|x_u \leq X_{(r)} \leq x_v)\right)$, the ordinary ratio strategy under the simple random sample design $(\bar{y}_{rs}, P_0(s))$ or under Lahiri–Miduno–Sen's (see Chap. 2) sampling design $(\bar{y}_{rs}, P_{LMS}(s))$ and the quantile ratio-type strategy $(\bar{y}_{rqs}, P_r(s))$.

The accuracy of the sampling strategies dependent on the sampling design $P_r(s|x_u \leq X_{(r)} \leq x_v)$ increases when the rank r of the order statistics $X_{(r)}$ increases and is best when $r = n$. The quantile ratio strategy $(\bar{y}_{rqs}, P_r(s))$ can be more accurate than other considered strategies for the small sample size when the outliers are not removed from the population. When the outliers are removed from the data the ratio strategies based on moments of the auxiliary variable are better than the quantile-dependent strategies. Moreover, the quantile strategies dependent on the conditional sampling design $P_r(s|x_u \leq X_{(r)} \leq x_{N-n+r})$ are better than those dependent only on the moments provided $r = n = 3$ and $u \leq 40$. Hence, in general, the accuracy of the quantile-type sampling strategies is comparable with accuracy of the ratio sampling strategies dependent on the moments of the auxiliary variables, but only in the case of small sample size.

The above analysis leads to the conclusion that, in a population with outliers the efficiency of the considered quantile-dependent ratio or regression estimation strategies is usually better than the efficiency of such type estimators, but only when the outliers are removed from the data. Let us note that the variable under study and the auxiliary variable are less correlated ($\rho = 0.97$) in the case when the outliers are removed from the data than in the case when they are not removed from the data ($\rho = 0.99$). Thus, these values could indicate that the efficiency of the estimation strategies depends on the correlation coefficient of the variables rather than on the existence of outliers in the data. Accordingly, a larger analysis of the dependence between the efficiency and the correlation of a variable under study and an auxiliary variable is needed. Moreover, the efficiency of the estimation strategies should be considered under the symmetrical distribution of the variable under study and the auxiliary variable.

We continue the analysis of the efficiency of the strategies on the basis of the methods described in the previous section with an artificial dataset of the size $N = 300$.

Table 4.2 The efficiency coefficients of the estimation strategies. The normal distribution $(X, Y) \sim N(10, 10, 1, 1, \rho)$. The size of the population is $N = 300$. The number of the iterations: 2000

Strategies	$\rho \backslash$ n:	3	6	9	15	30
$(\bar{y}_{rs}, P_{LMS}(s))$	0.5	99	99	101	95	99
	0.8	40	42	40	43	43
	0.95	11	11	11	10	10
$(\bar{y}_{regs}, P_{SS}(s))$	0.5	111	90	88	80	78
	0.8	52	48	45	44	44
	0.95	15	12	11	11	11
$(\bar{y}_{HTs}, P_{Sd}(s))$	0.5	113	108	108	109	111
	0.8	48	50	49	48	48
	0.95	11	10	11	10	11
$(\bar{y}_{rs}, P_0(s))$	0.5	97	102	96	96	97
	0.8	41	43	43	43	40
	0.95	13	12	13	12	12
$(\bar{y}_{regs}, P_0(s))$	0.5	562	107	91	83	79
	0.8	145	52	46	44	39
	0.95	47	14	12	11	11
$\left(\bar{y}_{regHTqs}, P_{n,1}^{(-)}(s)\right)$	0.5	682	129	110	93	86
	0.8	264	66	53	49	46
	0.95	13	13	13	13	13
$\left(\bar{y}_{regHTs}, P_{n,1}^{(-)}(s)\right)$	0.5	560	118	99	85	77
	0.8	248	57	46	43	43
	0.95	67	16	12	11	10
$(\bar{y}_{HTs}, P_n(s))$	0.5	77	81	90	94	95
	0.8	64	79	85	92	94
	0.95	46	71	82	91	97
$(\bar{y}_{rHTs}, P_n(s))$	0.5	97	96	95	98	99
	0.8	44	44	43	41	43
	0.95	11	11	11	10	11

Source own calculations

Three sets of data are generated according to the two-dimensional normal distributions with parameters $N(10, 10, 1, 1, \rho)$ for the correlation coefficients $\rho = 0.5$, $\rho = 0.8$ and $\rho = 0.95$. Next, three sets of data are generated according to the two-dimensional log-normal distributions $N(0, 0, 1, 1, \rho)$ for the correlation coefficients $\rho = 0.5$, $\rho = 0.8$, and $\rho = 0.95$. The accuracy of the strategies was assessed on the basis of computer simulation analysis. The samples were replicated at leas 1000 times. The relative efficiency coefficients are calculated according to the expression (4.1) and are shown in Tables 4.2 and 4.3.

Table 4.3 The efficiency coefficients of the estimation strategies. The log-normal distribution $(ln(X), ln(Y)) \sim N(0, 0, 1, 1, \rho)$. The size of the populations is $N = 300$. The number of the iterations: 2000

Strategies	$\rho \backslash$ n:	3	6	9	15	30
$(\bar{y}_{rs}, P_{LMS}(s))$	0.5	84	89	83	87	84
	0.8	44	40	41	46	43
	0.95	12	15	16	16	18
$(\bar{y}_{regs}, P_{SS}(s))$	0.5	110	93	85	80	78
	0.8	65	46	47	42	39
	0.95	17	15	15	18	18
$(\bar{y}_{HTs}, P_{Sd}(s))$	0.5	94	84	87	95	98
	0.8	24	24	23	23	22
	0.95	5	5	6	5	5
$(\bar{y}_{rs}, P_0(s))$	0.5	88	86	88	80	81
	0.8	43	44	43	43	43
	0.95	12	14	15	19	17
$(\bar{y}_{regs}, P_0(s))$	0.5	131	104	89	77	72
	0.8	181	59	45	43	41
	0.95	35	17	17	20	20
$\left(\bar{y}_{regHTqs}, P_{n,1}^{(-)}(s)\right)$	0.5	671	149	106	87	81
	0.8	358	85	58	50	46
	0.95	20	12	11	10	10
$\left(\bar{y}_{regHTs}, P_{n,1}^{(-)}(s)\right)$	0.5	628	141	98	85	79
	0.8	329	71	50	46	43
	0.95	18	12	11	10	10
$(\bar{y}_{HTs}, P_n(s))$	0.5	83	89	91	96	94
	0.8	70	84	91	92	93
	0.95	24	40	49	63	77
$(\bar{y}_{rHTs}, P_n(s))$	0.5	87	83	86	81	80
	0.8	45	43	45	42	45
	0.95	10	13	14	15	17

Source own calculations

The analysis of the data in the tables leads to the conclusion that the estimation strategies based on the moment-dependent sampling designs are usually slightly more efficient in the case of the estimation of the mean in the normal population than in the case of the estimation of the mean in the log-normal population. When the correlation coefficient between the variable under study and the auxiliary variable is equal to 0.5, the efficiency coefficients of all considered strategies are approximately equal to 100% or greater than 100%.

In the case of the normal distributions of the variable under study and of the auxiliary variable, the analysis of data in Table 4.2 allows us to draw the following conclusions. The strategy $(\bar{y}_{HTs}, P_n(s))$ is the worst among the considered nine estimation strategies for the data examined. In the case when the correlation coefficient is equal to $\rho = 0.5$, the efficiencies of all the estimation strategies are close to 100% or greater than 100%. When $\rho = 0.8$ ($\rho = 0.95$) and $n \geq 6$, the estimation strategies are approximately two (ten) times more accurate than the simple random sample mean. If $n \geq 6$ and $\rho \geq 0.8$ the accuracies of the estimators $(\bar{y}_{rs}, P_{LMS}(s))$, $(\bar{y}_{HTs}, P_{Sd}(s))$, and $(\bar{y}_{rHTs}, P_n(s))$ are similar each to other and not worse than the accuracies of the remaining strategies.

When the log-normal population is considered, the data in Table 4.3 lead to the following conclusions. Provided the sample size is not less than 6, the values of the relative efficiency coefficients do not differ significantly from each other, except for the efficiency coefficients of the Horvitz–Thompson strategy $(\bar{y}_{HTs}, P_n(s))$. When $\rho = 0.8$ or $\rho = 0.95$, the strategy $(\bar{y}_{HTs}, P_{Sd}(s))$ is the most efficient among the strategies considered. When $n \geq 6$ and $\rho = 0.5$ the efficiency coefficients are not less than 72%. In the case when $n \geq 6$ and $\rho = 0.8$, the strategies are approximately two times more efficient than the simple random sample mean, except for the strategy $(\bar{y}_{HTs}, P_n(s))$, for which the efficiency coefficient is not less than 84%.

4.3 Efficiency of Estimation Strategies Dependent on the Sum of Order Statistics

Based on the results from the previous section, we conclude that in the case of high correlation between the variable under study and the auxiliary variable, some of the strategies considered are significantly profitable in the sense of increasing efficiency estimation. That is why our analysis is limited to the case when the correlation coefficient between the variable under study and the auxiliary variable has the value 0.95. Moreover, we assume that the sample size is equal to 6.

The Horvitz–Thompson-type ratio or regression estimators under the conditional sampling design proportional to the sum of order statistics of the auxiliary variables are considered in Tables 4.4 and 4.5, where, except for the relative efficiency coefficients of the strategies, we can find the correlation coefficients between inclusion probabilities and the auxiliary variable for the cases when $X_{(r)} + X_{(u)} > 2c$ where $c = 0$, $c = \bar{x}$, $c = q_{0.5}$, or $c = q_{0.7}$ and $q_\alpha, 0 < \alpha < 1$ is the population quantile of order α of the auxiliary variable. Based on Table 4.4, we infer that in the case of the normally distributed data, the efficiency coefficients have values from the interval [9%; 27%]. When data have a log-normal distribution, the data, as seen in Table 4.5, let us conclude that the efficiency coefficients have values from the interval [5%; 10%]. Thus, when the variable under study and the auxiliary variables have a joint log-normal distribution, the strategies are slightly more efficient than in the case when those variables are normally distributed. Moreover, the variabil-

Table 4.4 The relative efficiency coefficients of the strategies $P_{r,u}^{(+)}(s|X_{(r)} + X_{(u)} > 2c)$ for $n = 6$ and the correlation coefficient $\rho(x, \pi)$. The normal distribution $(X, Y) \sim N(10, 10, 1, 1, \rho)$, $\rho = 0.95$, $N = 300$. The number of the iteration: 1000

r	u	$c = 0$			$c = q_{0.5}$			$c = \bar{x}$			$c = q_{0.7}$		
		$\bar{y}_{regHTs}$	$\bar{y}_{rHTs}$	$\rho(x,\pi)$	$\bar{y}_{regHTs}$	$\bar{y}_{rHTs}$	$\rho(x,\pi)$	$\bar{y}_{regHTs}$	$\bar{y}_{rHTs}$	$\rho(x,\pi)$	$\bar{y}_{regHTs}$	$\bar{y}_{rHTs}$	$\rho(x,\pi)$
1	2	11	10	0.78	17	16	0.88	15	15	0.88	14	14	0.88_0
1	3	14	11	0.87	13	14	0.94	14	14	0.94	19	19	0.91_0
1	4	13	11	0.93	14	14	0.97	13	13	0.97	15	15	0.88_0
1	5	13	11	0.97	13	13	0.97	13	12	0.97	13	13	0.90_0
1	6	12	10	0.93	14	12	0.92	13	11	0.92	12	12	0.83_0
2	3	13	11	0.91	20	16	0.88	20	16	0.88	27	24	0.86
2	4	12	10	0.96	15	12	0.94	16	14	0.94	21	18	0.91
2	5	13	11	0.99	13	11	0.97	13	12	0.97	16	13	0.92
2	6	13	10	0.96	12	11	0.98	11	11	0.99	12	11	0.89
3	4	13	12	0.95	14	12	0.89	13	11	0.89	19	16	0.86
3	5	13	10	0.97	11	10	0.94	11	11	0.94	15	14	0.91
3	6	12	10	0.95	9	9	0.98	11	11	0.98	12	11	0.92
4	5	13	11	0.94	11	10	0.90	11	11	0.90	13	12	0.87
4	6	13	11	0.91	11	11	0.96	11	11	0.96	12	11	0.92
5	6	12	10	0.85	11	10	0.95	11	11	0.95	12	11	0.88

Source own calculations

Table 4.5 The relative efficiency coefficients of the strategies $P_{r,u}^{(+)}(s|X_{(r)} + X_{(u)} > 2c)$ for $n = 6$ and the correlation coefficient $\rho(x, \pi)$. The log-normal distribution $(ln(X), ln(Y)) \sim N(0, 0, 1, 1, \rho)$, $\rho = 0.95$, $N = 300$. The number of the iteration is 1000

		$c = 0$			$c = q_{0.5}$			$c = \bar{x}$			$c = q_{0.7}$		
r	u	$\bar{y}_{regHTs}$	$\bar{y}_{rHTs}$	$\rho(x,\pi)$	$\bar{y}_{regHTs}$	$\bar{y}_{rHTs}$	$\rho(x,\pi)$	$\bar{y}_{regHTs}$	$\bar{y}_{rHTs}$	$\rho(x,\pi)$	$\bar{y}_{regHTs}$	$\bar{y}_{rHTs}$	$\rho(x,\pi)$
1	2	8	7	0.68	9	10	0.66	6	7	0.65	10	7	0.65
1	3	9	7	0.81	6	6	0.79	6	6	0.79	6	6	0.79
1	4	9	8	0.87	6	6	0.84	6	5	0.84	6	5	0.84
1	5	9	10	0.86	7	6	0.83	6	6	0.82	6	5	0.82
1	6	8	7	0.72	9	8	0.71	8	7	0.70	8	7	0.70
2	3	9	7	0.83	8	7	0.82	7	6	0.83	7	6	0.81
2	4	9	7	0.90	7	7	0.89	6	5	0.89	6	5	0.89
2	5	8	7	0.96	7	6	0.95	6	6	0.95	6	6	0.95
2	6	9	8	0.84	8	7	0.83	8	7	0.82	8	6	0.81
3	4	9	7	0.85	7	7	0.85	6	6	0.85	7	5	0.85
3	5	9	8	0.92	7	6	0.91	6	6	0.91	6	6	0.91
3	6	9	7	0.88	8	7	0.86	8	7	0.85	7	6	0.85
4	5	9	8	0.85	8	7	0.84	7	7	0.84	7	6	0.84
4	6	9	7	0.83	8	7	0.82	8	7	0.80	8	7	0.80
5	6	9	7	0.70	7	7	0.68	8	7	0.67	8	7	0.67

Source own calculations

ity of the efficiency coefficient in the case of the log-normally distributed data is less than in the case of the normally distributed data. When the data are normally distributed, the conditional regression and the ratio strategies, under the assumption that $X_{(3)} + X_{(6)} > 2q_{0.5}$, are the most accurate because their efficiency coefficients are equal to 9%. If the data are log-normally distributed, then the efficiency coefficients of the four conditional ratio strategies are lowest and equal to 5% as shown in Table 4.5. One of them is the strategy $\left(\bar{y}_{rHTs}, P_{2,4}^{(+)}(s|X_{(r)} + X_{(u)} > 2\bar{x})\right)$. Hence, in general, we can conclude that the conditional estimation strategies can be more efficient than their unconditional versions.

Moreover, the analysis of the data in Tables 4.4, 4.5 leads to the following conclusions. In the case of the normal population, the efficiency of the ratio-type strategies based on sampling designs $P_{r,u}^{(+)}(s)$ are approximately equal to the efficiency of the Lahiri–Midzuno–Sen strategy (considered in Chap. 3) $(\bar{y}_{rs}, P_{LMS}(s))$. In the case of the log-normal population, the efficiency coefficients of the ratio-type strategies under the sampling design $P_{r,u}^{(+)}(s|X_{(r)} + X_{(u)} > 2c)$ have values from the interval [5%; 10%] whereas the efficiency coefficients of the moment-dependent strategies have values from the interval [5%; 17%]. Particularly,

$$deff(\bar{y}_{HTs}, P_{Sd}(s)) = 5\% = deff(\bar{y}_{rHTs}, P_{2,4}^{(+)}(s|X_{(2)} + X_{(4)} > 2\bar{x})) = 5\%.$$

Thus, the conditional ratio-type strategy dependent on the sum of the two order statistics and the unconditional Sampford's sampling strategy are equally efficient.

The values of the correlation coefficient $\rho(x, \pi)$ depends on the ranks r and u of the order statistic of the auxiliary variable. For instance, the shapes of the dependence between the values of the auxiliary variable and the inclusion probabilities are presented by Figs. 4.1, 4.2, 4.3 for the case of the normal population. When $r = 2$, $u = 5$, and $n = 6$, the relationship between the auxiliary variable and the inclusion probabilities is close to linear because $\rho(x, \pi) = 0.995$ (Fig. 4.2). Hence, in the case of the sample size $n = 6$, the inclusion probabilities of the sampling design $P_{2,5}^{(+)}(s)$

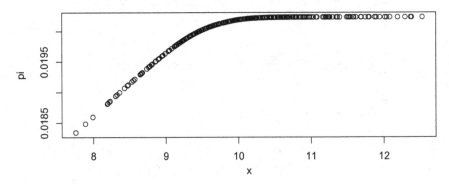

Fig. 4.1 The inclusion probabilities of the sampling design $P_{1,2}^{(+)}(s)$, $n = 6$, $\rho(x, \pi) = 0.787$, $X \sim N(10, 1)$. *Source* own preparation

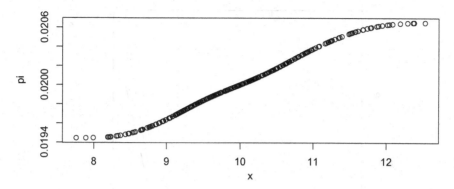

Fig. 4.2 Inclusion probabilities for $P_{2,5}^{(+)}(s)$, and $n = 6$, $\rho(x, \pi) = 0.995$, $X \sim N(10, 1)$. *Source* own preparation

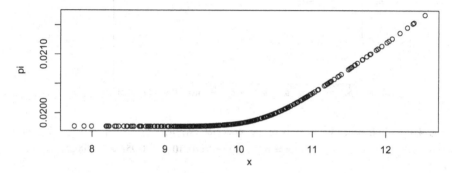

Fig. 4.3 The inclusion probabilities for $P_{5,6}^{(+)}(s)$, and $n = 6$, $\rho(x, \pi) = 0.849$, $X \sim N(10, 1)$. *Source* own preparation

are almost proportional to the appropriate values of the auxiliary variable. Let us note that the sampling designs with such inclusion probabilities are expected in survey sampling applied, e.g. in accounting. If $r = 1$ and $u = 2$ or $r = 5$ and $u = 6$, the relationship between the inclusion probabilities and the auxiliary variable is not linear, as shown by Figs. 4.1 and 4.3. In these cases, the correlation coefficients between the auxiliary variable, and the inclusion probabilities are not greater than the value 0.849.

The analysis of the data in Table 4.5 indicates that in the case of the log-normal data, the estimation strategies using the conditional versions of the sampling design $P_{r,u}^{(+)}(s|X_{(r)} + X_{(u)} > 2c)$ are usually more efficient then in the case of their unconditional versions. For instance, in the case of the regression Horvitz–Thompson strategy, we have

$$deff(\bar{y}_{regHTs}, P_{1,5}^{(+)}(s)) = 10\% > deff(\bar{y}_{regHTs}, P_{1,5}^{(+)}(s|X_{(r)} + X_{(u)} > 2q_{0.7}) = 6\%.$$

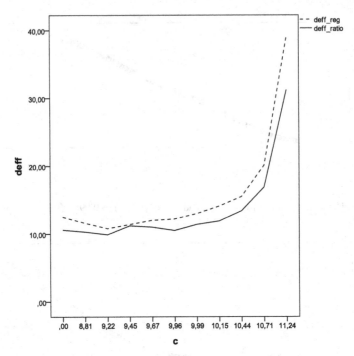

Fig. 4.4 The efficiency coefficients of the strategies $(\bar{y}_{regHTs}, P_{2,5}^{(+)}(s|X_{(r)} + X_{(u)} > 2c)$ and $(\bar{y}_{rHTs}, P_{2,5}^{(+)}(s|X_{(r)} + X_{(u)} > 2c), n = 6, (X, Y) \sim N(10, 10, 1, 1, 0.95), N = 300$. *Source* own preparation

Based on the analysis of the data in Table 4.4, we infer that in the case of the normal distribution of the variables and all combinations of the ranks r and u of the order statistic, the efficiencies of the unconditional strategies are similar. However, the strategies with the conditional versions of the sampling designs are usually more efficient for large ranks r and u than for small ranks r and u.

The analysis of the data in Table 4.4 allows us to say that in the case of the normal distribution of the data, the efficiency coefficients of the regression (ratio) strategies have values from the interval [11%; 14%] ([10%; 12%]) for $c = 0$, and they are in the interval [12%; 27%] ([11%;19%]) for $c = q_{0.7}$. In particular, in this case, the analysis of the data in Fig. 4.4 indicates that the conditional regression estimation strategy $(\bar{y}_{regHTs}, P_{2,5}^{(+)}(s|X_{(2)} + X_{(5)} > 2c))$ and the ratio strategy $(\bar{y}_{rHTs}, P_{2,5}^{(+)}(s|X_{(2)} + X_{(5)} > 2c))$ for $c > 0$ are not more efficient than their unconditional versions for $c = 0$. Similarly, the analysis of Table 4.5 data allows us to conclude that in the case of the log-normal distribution of the data the efficiency coefficients of the regression (ratio) strategies have values from the interval [8%; 9%] ([7%; 10%]) for $c = q_{0.5}$. Moreover, in this case, based on the Fig. 4.5 data, we can say that the there is a tendency of decrease in the efficiency coefficients of the conditional strategies $(\bar{y}_{regHTs}, P_{2,5}^{(+)}(s|X_{(2)} + X_{(5)} > 2c))$ and

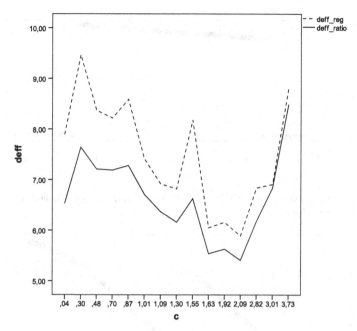

Fig. 4.5 The efficiency coefficients of the strategies $(\bar{y}_{regHTs}, P_{2,5}^{(+)}(s|X_{(r)} + X_{(u)} > 2c)$ and $(\bar{y}_{rHTs}, P_{2,5}^{(+)}(s|X_{(r)} + X_{(u)} > 2c), n = 6, (ln(X), ln(Y)) \sim N(0, 0, 1, 1, 0.95), N = 300$. *Source* own preparation

$(\bar{y}_{rHTs}, P_{2,5}^{(+)}(s|X_{(2)} + X_{(5)} > 2c))$ until $c = 2.09$, and then, the efficiency coefficients increase. Thus, in the case of the log-normal distribution of the data, the conditional strategies are more efficient than their unconditional versions.

From a practical point of view, the inclusion probabilities of the conditional sampling design is calculated by a computer more swiftly than its unconditional version. For instance, the inclusion probabilities of the sampling design $P_{2,5}^{(+)}(s)$ are calculated to be 2.5 times larger than in the case $P_{2,5}^{(+)}(s|X_{(2)} + X_{(5)} > 2q_{0.7})$. The correlation coefficients between the inclusion probabilities and the auxiliary variable of both strategies are approximately the same. Moreover, the efficiencies of the conditional strategies are slightly better than their unconditional versions (see Table 4.5).

Now, let us consider the following conditional strategies (inclusion probabilities are shown by Figs. 4.6, 4.7, 4.8, 4.9). The relationship between the auxiliary variable and the inclusion probabilities of the sampling design proportional to the sum of the order statistics $X_{(2)} + X_{(5)}$, under the condition that $X_{(2)}$ is not less than $q_{0.1}$ and $X_{(5)}$ is not greater than $q_{0.9}$, is shown by Fig. 4.6. In this case, the population elements indexed by small or large values of the auxiliary variable placed outside the interval $[q_{0.1}, q_{0.9}]$ are drawn to the sample with lower probabilities than the elements identified by the auxiliary variable values from this interval. In the case of the conditional sampling design under consideration, the efficiency coefficients of

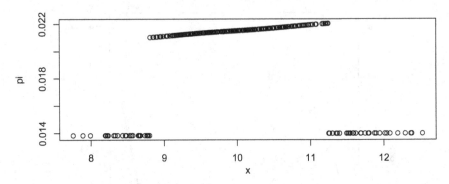

Fig. 4.6 The inclusion probabilities for $P_{2,5}^{(+)}(s|X_{(2)} \geq q_{0.1}, X_{(5)} \leq q_{0.9})$, and $n = 6$. $X \sim N(10, 1)$. *Source* own preparation

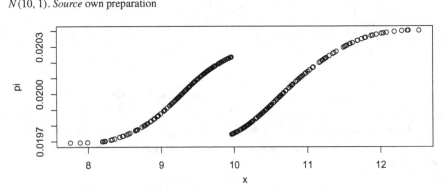

Fig. 4.7 The inclusion probabilities for $P_{2,5}^{(+)}(s|X_{(2)} \leq q_{0.5} < X_{(5)})$, and $n = 6$. $X \sim N(10, 1)$. *Source* own preparation

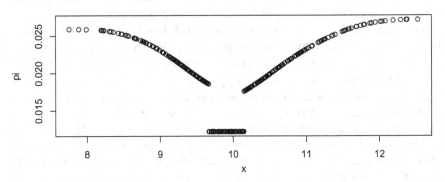

Fig. 4.8 The inclusion probabilities for $P_{2,5}^{(+)}(s|X_{(2)} < q_{0.4}, X_{(5)} \geq q_{0.6})$, and $n = 6$. $X \sim N(10, 1)$. *Source* own preparation

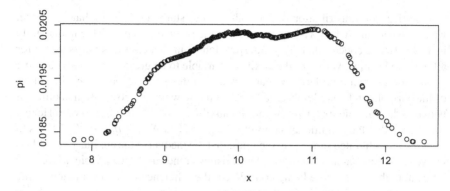

Fig. 4.9 The inclusion probabilities for $P_{2,5}\left(s\middle|\left|\frac{X_{(2)}+X_{(5)}}{2}-\bar{x}\right|\leq 0.844\right)$, and $n=6$. $X \sim N(10, 1)$. *Source* own preparation

the regression and ratio estimators are equal to 13% and 10%, respectively. In the case of the unconditional version of the sampling design, those coefficients are equal to 13% and 11%, respectively. Thus, both versions of the strategies have similar efficiencies. From a practical point of view, that property of the sampling design can be useful, such as when a survey statistician expects non-response errors for small or large values of an auxiliary variable. The sampling design considered prefers not to draw the population elements indexed by the outliers of an auxiliary variable to the sample. Moreover, this conditional sampling design can be useful when samples are drawn from the domains of a population identified by auxiliary variable observations from a pre-assigned interval. See the example in the next section.

Figure 4.7 shows the relationship between the auxiliary variable and the inclusion probabilities of the sampling design proportional to the sum of the order statistics $X_{(2)} + X_{(5)}$ under the assumption that values of these order statistics are separated by the population median $q_{0.5}$ of the auxiliary variable. In the case of the normal population, we have the following:

$$deff(\bar{y}_{regHTs}, P_{2,5}^{(+)}(s|X_{(2)} \leq q_{0.5} < X_{(5)})) = 11\% < deff(\bar{y}_{regHTs}, P_{2,5}^{(+)}(s)) = 13\%$$

$$deff(\bar{y}_{rHTs}, P_{2,5}^{(+)}(s|X_{(2)} \leq q_{0.5} < X_{(5)})) = deff(\bar{y}_{rHTs}, P_{2,5}^{(+)}(s)) = 11\%$$

Thus, the conditional version of the regression strategy is more efficient than its unconditional version.

The observation in the sample s data $((X_r, Y_r), r = 1, ..., n)$ ordered by the values of the auxiliary variable can be denoted by $((X_{(r)}, Y_{[r]}), r = 1, ..., n)$ where $X_{(r)}$ is the rth-order statistic and $Y_{[r]}$ is concomitant with $X_{(r)}$. The assumption $X_{(2)} \leq q_{0.5} < X_{(5)}$ leads to the reduction of the entire sample space $\mathbf{S}$ to the subspace $\mathbf{S}_*$ including only such samples that $x_{(2)} \leq q_{0.5} < x_{(5)}$ where $x_{(2)}$ and $x_{(5)}$ are values of the auxiliary variable order statistics $X_{(2)}$ and $X_{(5)}$ observed in the sample. In the case of the symmetrical joint distribution of a variable under study and the auxiliary variable, we can expect that when $X_{(2)} \leq q_{0.5} < X_{(5)}$, the median of the variable under study is also between concomitant $Y_{[2]}$ and $Y_{[5]}$. Hence, in this case it is not possible for all observed sample values of the auxiliary variable to be greater or smaller than the population median. Similarly, we can expect that this property addresses the values of the variable under study observed in the sample when the correlation coefficient between the variable under study and the auxiliary variable is close to one. Hence, we can say that the conditional sampling design $P_{2,5}^{(+)}(s|X_{(2)} \leq q_{0.5} < X_{(5)})$ leads to drawing the calibrated samples in the sense that the median of the auxiliary variable is between the second- and the fifth-order statistics of the auxiliary variable. Moreover, let us note that in the case being considered, the efficiency of the estimation of the mean does not decreases, when the sample space $\mathbf{S}$ is reduced to its subspace $\mathbf{S}_*$ in the conditional sampling design.

Figure 4.8 shows the relationship between the values of the auxiliary variable and the inclusion probabilities of the conditional sampling design $P_{2,5}^{(+)}(s|X_{(2)} < q_{0.4}, X_{(5)} \geq q_{0.6})$. The design prefers drawing to the sample the population elements identified by auxiliary variable values less than the quantile of rank 0.4 or greater than the quantile of rank 0.6. Hence, the conditional sampling design type can be useful when samples are drawn from domains of a population. For instance, let the elements of the first domain be identified by the values of the auxiliary variable less than the quantile $q_{0.4}$. The elements of the second domain are indexed by the values of the auxiliary variable that lay between the quantiles $q_{0.4}$ and $q_{0.6}$, and the last domain elements are identified by the values of the auxiliary variable greater than $q_{0.6}$. Thus, the just-defined conditional sampling design prefers drawing population elements from the first and last domains, as seen in the results in Fig. 4.8.

The inclusion probabilities of the naturally constructed sampling design are shown in Fig. 4.9. In this case, the conditional sampling design prefers to draw the population elements to the sample in such the way that the mean of the order statistics $X_{(2)}$ and $X_{(5)}$ differs from the population mean of the auxiliary variable by not more than the pre-assigned constant c. In the case of the normal population, we have the following:

$$deff\left(\bar{y}_{regHTs}, P_{2,5}\left(s|\,|\bar{X}_{(2,5)} - \bar{x}| \leq c\right)\right) = 11\% < deff\left(\bar{y}_{regHTs}, P_{2,5}(s)\right) = 13\%$$

$$deff\left(\bar{y}_{rHTs}, P_{2,5}\left(s|\,|\bar{X}_{(2,5)} - \bar{x}| \leq c\right)\right) = 10\% < deff\left(\bar{y}_{rHTs}, P_{2,5}(s)\right) = 11\%$$

where

$$\bar{X}_{(2,5)} = \frac{X_{(2)} + X_{(5)}}{2}$$

and $c = z_{0.9} * \sqrt{v(x)/2} = 0.884$ where $z_{0.9}$ is the quantil of the rank 0.9 of the standard normal variable.

Let us note that the statistic $\bar{X}_{(2,5)}$ can be treated as the estimator of the population mean $\bar{x}$. Moreover, when a value of $\bar{X}_{(2,5)}$ is close to $\bar{x}$, all the observed sample values of the auxiliary variable are close to the mean $\bar{x}$ except the values of the first- and the last- order statistics denoted by $X_{(1)}$ and $X_{(n)}$, respectively. Thus, we can conclude that the conditional sampling design $P_{2,5}\left(s \mid \left|\frac{X_{(2)}+X_{(5)}}{2} - \bar{x}\right| \leq 0.844\right)$ prefers to draw the population elements indexed by the auxiliary variable observations close to its mean value. The sample being selected in such a way allows us to expect that it includes the values of the variable under study close to its mean value because of the high correlation coefficient between the auxiliary variable and the variable under study. That is why in the case considered, the conditional sampling strategies improve the accuracy of the estimation of the population mean.

4.4 Efficiency Estimation of Domain Mean

Let the population U be divided into three disjoint domains denoted by U_h where $h = 1, 2, 3$ and $U = \bigcup_{h=1}^{3} U_h$. In each domain, a two-dimensional variable is defined. Its values are outcomes generated from the normally distributed random variable with the pre-assigned parameters. The purpose is the estimation of the domain's mean of the variable under study on the basis of the sample selected from the whole population. The considered below estimators of an hth domain mean are calculated on the basis of the sub-sample s_h consisted only of elements of the hth domain. So, $s = s_1 \cup s_2 \cup s_3$ where s is drawn from the whole population according to a sampling design $P(s)$. The size of the sub-sample s_h is denoted by n_h. We assume that $n_h > 1$ for $h = 1, 2, 3$ and $n_1 + n_2 + n_3 = n$.

We consider the following four direct estimation strategies. The first one is based on the simple random sample selected without replacement from the whole population. It is the following mean of the values of the variable under study observed in the sub-samples:

$$\bar{y}_{s_h} = \frac{1}{n_h} \sum_{k \in s_h} y_k. \tag{4.2}$$

The second strategy involves the Horvitz–Thompson-type statistics

$$\tilde{y}_{HTs_h} = \frac{1}{N_{HTs_h}} \sum_{k \in s_h} \frac{y_k}{\pi_k}, \qquad \text{where} \qquad N_{HTs_h} = \sum_{k \in s_h} \frac{1}{\pi_k}. \tag{4.3}$$

The third strategy is based on the ordinary regression estimator:

$$\bar{y}_{regs_h} = \bar{y}_{s_h} + a_{s_h}(\bar{x}_h - \bar{x}_{s_h}) \tag{4.4}$$

where
$$a_{s_h} = \frac{v_{s_h}(x, y)}{v_{s_h}(x)},$$

$$v_{s_h}(x, y) = \frac{1}{n_h - 1} \sum_{k \in s_h}(x_k - \bar{x}_{s_h})(y_k - \bar{y}_{s_h}), \qquad v_{s_h}(x) = v_{s_h}(x, x),$$

The last strategy is based on the regression Horvitz–Thompson estimator

$$\tilde{y}_{regHTs_h} = \tilde{y}_{HTs_h} + \tilde{a}_{HTs_h}(\tilde{x}_h - \tilde{x}_{HTs_h}) \tag{4.5}$$

where
$$\tilde{a}_{HTs_h} = \frac{\tilde{v}_{HTs_h}(x, y)}{\tilde{v}_{HTs_h}(x)},$$

$$\tilde{v}_{HTs_h}(x, y) = \frac{1}{N_{HTs_h} - 1} \sum_{k \in s_h} \frac{(x_k - \tilde{x}_{HTs_h})(y_k - \tilde{y}_{HTs_h})}{\pi_k}, \qquad \tilde{v}_{HTs_h}(x) = \tilde{v}_{HTs_h}(x, x).$$

The comparisons of the accuracy of strategies are based on the following relative efficiency coefficients:
$$deff_h = \frac{MSE\left(t_{s_h}, P(s)\right)}{V\left(\bar{y}_{s_h}, P_0(s)\right)}.$$

We assume that in the population of the size $N = 600$, there are three domains. Each of them is of the size $N_1 = N_2 = N_3 = 200$. The data in the first domain are the values generated from the two-dimensional probability distribution with parameters $N(10, 10, 1, -0.95)$. The data in the second and the third domains are the values generated from the distributions $N(11, 11, 1, 1, 0)$ and $N(13, 13, 1, 1, 0.95)$, respectively. Hence, the data are generated according to the mixture of the three two-dimensional normal distributions. The scatter-plot of the data is shown by Fig. 4.10. The accuracy analysis of the strategies is based on the simulation experiment. The sample s is generated according to the assumed sampling scheme. Next, the sub-samples s_h, $h = 1, 2, 3$ are identified in the sample s. If the size of at least one sub-sample is less than 2, then the sample s is rejected and a new one is drawn from the population. When the size of the each sub-sample is greater or equal to 2, then the values of estimators are calculated on the basis of the data observed in the sub-samples. Next, the new sample is generated according the same sampling scheme. This procedure is replicated 2000 times. Finally, the defined above efficiency coefficients are calculated.

The efficiency coefficients of the Horvitz–Thompson estimator-type strategy are not less than 100%. That is why they are not taken into account in Table 4.6. The analysis of the data in Table 4.6 leads to the following conclusions. In general, the efficiency coefficients of all regression-type estimators increases when the sample size increases, except in the case of the third domain where efficiency coefficients

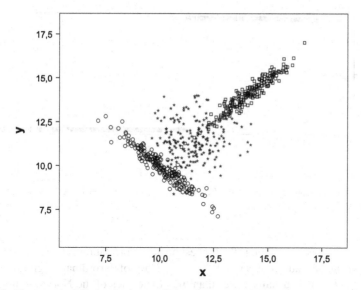

Fig. 4.10 The spread of the data in the population. *Source* own preparation

Table 4.6 The relative efficiencies of the regression and the ratio estimators for the domains no. $h = 1, 2, 3$ and the sample sizes: $n = 12, ..., 60$. The normal distribution mixture. The population size is $N = 600$

Strategy	h/n:	12	18	24	30	60
$(\bar{y}_{regs_d}, P_0(s))$	1	6	15	29	51	169
	2	3	5	10	16	49
	3	1	0	0	0	1
$(\bar{y}_{regHTs_d}, P_n(s))$	1	6	10	15	21	52
	2	3	4	6	9	21
	3	0	0	0	0	1
$(\bar{y}_{regHTs_d}, P_{2,n-1}^{(+)}(s))$	1	5	11	16	23	54
	2	2	5	7	10	21
	3	0	0	0	0	1
$(\bar{y}_{regHTs_d}, P_{2,n-1}^{(+)}(s\|X_{(5)} \leq q_{0.5}))$	1	3	6	11	15	68
	2	4	7	14	22	100
	3	0	1	2	4	16
$(\bar{y}_{regHTs_d}, P_{n-1,2}^{(-)}(s))$	1	5	10	15	21	53
	2	2	5	7	9	20
	3	0	0	0	0	1

Source calculations

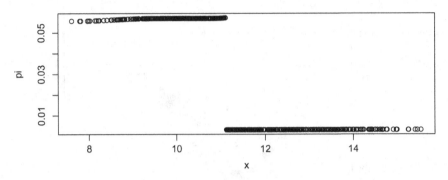

Fig. 4.11 The inclusion probabilities for $P_{2,5}^{(+)}(s|X_{(5)} \leq q_{0.5})$, and $n = 18$. $X \sim N(10, 1)$. *Source own preparation*

are close to 1% because there is the highest positive correlation between the variable under study and the auxiliary variable. The efficiency of the ordinary regression strategy ($\bar{y}_{regs_d}$, $P_0(s)$) is usually lower than the efficiencies of the Horvitz–Thompson regression-type strategies under the sampling designs dependent on order statistics. In general, the analysis allows us to conclude that Horvitz–Thompson's regression strategy based on unconditional sampling designs dependent on the order statistics are not less efficient than the ordinary ratio and regression strategies.

Let us suppose that we are especially interested in the estimation of the mean in the first domain with high accuracy. We assume that we know in advance that the first domain is approximately identified by the auxiliary variable values not greater than its median $q_{0.5} = 11.105$. In this case, we can prefer the conditional sampling design $P_{2,n-1}^{(+)}(s|X_{(5)} \leq q_{0.5})$ because the design prefers drawing population elements from a first domain (Figs. 4.10 and 4.11). On the basis of Table 4.6, data we can say that the considered conditional regression strategy improves the accuracy of the estimation of the mean in the first domain for the considered sample sizes except for the size of 30.

4.5 Estimation of Quantiles

Quantiles are very important parameters of a population. Let $q_{y,p}$ be the pth population quantile of a variable. It is well known (see, e.g. Rao et al. 1990 or Särndal et al. 1992, p. 199) that the estimator of pth quantile of a variable under study is as follows:

$$q_{y,p,s} = F_{yHTs}^{-1}(p) \tag{4.6}$$

where $F_{yHTs}^{-1}(p) = inf\{y_k \in s : F_{HTs}(y_k) \geq p\}$ and the estimator of the distribution function is

$$F_{HTs}(y) = \left(\sum_{k \in M_{y,s}} \frac{1}{\pi_k} \right) \left(\sum_{k \in s} \frac{1}{\pi_k} \right)^{-1},$$

$M_{s,y} = \{k : k \in s, \ y_k \leq y\}$ is the set of sample elements with values $y_k \leq y$. Particularly, in the case of the simple random sample drawn without replacement, we have $F_{HTs}(y) = \frac{m_{y,s}}{n}$ where $m_{y,s}$ is the size of the set $M_{y,s}$.

Instead of the statistic $q_{p,s}$, special ratio- or regression-type estimators are constructed to estimate the quantile. One of them is as follows, see, e.g. Arcos et al. (2007):

$$\hat{q}_{y,p,s} = q_{y,p,s} \frac{q_{x,p}}{q_{x,p,s}} \tag{4.7}$$

where $q_{x,p}$ is the pth population quantile of the auxiliary variable and $q_{x,p,s} = F_{xHTs}^{-1}(p)$ is the pth sample quantile of the auxiliary variable.

Now, we can look for the sampling design that leads to a more accurate estimation of the quantile. In Sect. 4.6, the following sampling design is considered:

$$P_{2,n-1}^{(+)} \left(s | X_{(2)} \geq c_1, X_{(n-1)} \leq c_2 \right)$$

Based on Fig. 4.6, we expect that it is reasonable to choose c_1 and c_2 in such a way that $c_1 \leq q_{x,p} \leq c_2$. To simplify the next analysis, let us assume that $c_1 = q_{x,p} - d_x/2$ and $c_2 = q_{x,p} + d_x/2$ where d_x is the population standard deviation of the auxiliary variable. Hence, the above sampling design takes the following form: $P_{2,5}^{(+)} \left(s | w_1 \left(q_{x,p}, \frac{d_x}{2} \right) \right)$ where

$$w_1 \left(q_{x,p}, \frac{d_x}{2} \right) = \left(s : X_{(2)} \geq q_{x,p} - \frac{d_x}{2}, X_{(5)} \leq q_{x,p} + \frac{d_x}{2} \right).$$

Fig. 4.9 let us construct the sampling design $P_{2,5}^{(+)} \left(s | w_2 \left(q_{x,p}, \frac{d_x}{2} \right) \right)$ where

$$w_2 \left(q_{x,p}, \frac{d_x}{2} \right) = \left(s : \left| \frac{X_{(2)} + X_{(5)}}{2} - q_{x,p} \right| \leq \frac{d_x}{2} \right).$$

The examples of the inclusion probabilities evaluated under the above circumstances are presented by Figs. 4.12, 4.13, 4.14, 4.15, 4.16, 4.17, 4.18, 4.19, 4.20.

Let us note that in the case of the sampling design $P_{2,5}^{(+)}(s | w_1(q_{x,0.1}, d_x/2))$, the maximum value of the inclusion probability is approximately eight times larger than the minimum value, see Fig. 4.12.

Under this sampling design and the simple random sample, the simulation analysis of the accuracy of the estimators $q_{y,p,s}$ and $\hat{q}_{y,p,s}$ has been processed. The results of the simulation are shown in Table 4.7 where the coefficients of the efficiency are evaluated based on the following ratio:

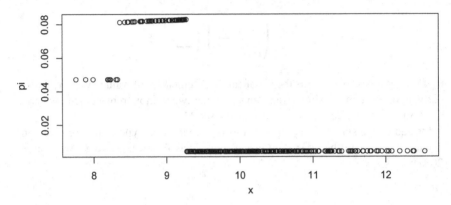

Fig. 4.12 The inclusion probabilities for $P_{2,5}^{(+)}(s|w_1(q_{x,0.1}, d_x/2))$, $q_{x,0.1} = 8.810$, $d_x = 0.932$ and $n = 6$. $X \sim N(10, 1)$. *Source* own preparation

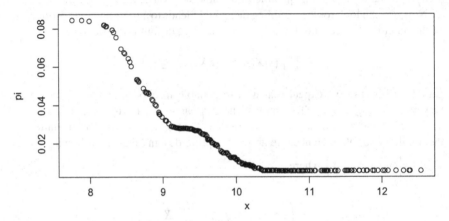

Fig. 4.13 The inclusion probabilities for $P_{2,5}^{(+)}(s|w_2(q_{x,0.1}, d_x/2))$, $q_{x,0.1} = 8.810$, $d_x = 0.932$, and $n = 6$. $X \sim N(10, 1)$. *Source* own preparation

$$e_p = \frac{mse(t_{y,p,s}, P(s|w_i(c)))}{mse(q_{y,p,s}, P_0(s))} 100\%, \quad i = 1, 2$$

where $mse(q_{y,p,s}, P(s|w_i(c)))$ is the mean square error of the estimation strategy of the quantile $q_{y,p}$ and $mse(q_{y,p,s}, P_0(s))$ is the estimator of the quantile under the simple random sample drawn without replacement, shown by the expression (4.6). The relative bias of the quantile estimation is determined by the equation:

$$b_p = \frac{(E(t_{y,p,s}, P(s|w_i(c))) - q_{y,p})^2}{mse(t_{y,p,s}, P_0(s))} 100\%$$

Table 4.7 shows the defined efficiency coefficients and relative bias of the estimation strategies under consideration evaluated on the basis of the computer simulation

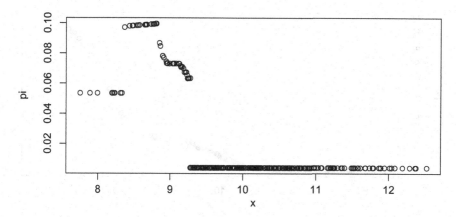

Fig. 4.14 The inclusion probabilities for $P_{2,5}^{(+)}(s|w_1(q_{x,0.1}, d_x/2), w_2(q_{x,0.1}, d_x/4))$, $q_{x,0.1} = 8.810$, $d_x = 0.932$, and $n = 6$. $X \sim N(10, 1)$. *Source* own preparation

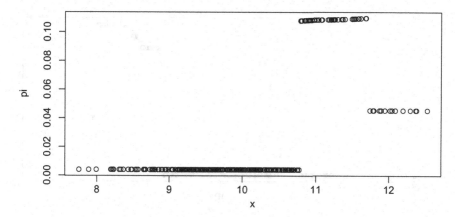

Fig. 4.15 The inclusion probabilities for $P_{2,5}^{(+)}(s|w_1(q_{x,0.9}, d_x/2))$, $q_{x,0.9} = 11.242$, $d_x = 0.932$, and $n = 6$. $X \sim N(10, 1)$. *Source* own preparation

analysis through replication of drawing the samples of the size $n = 6$ from the normal population of the size $N = 300$ with correlation coefficient $\rho = 0.95$ described in the Sect. 4.1.

In the analysis of Table 4.7, lets us conclude that, under high correlation between the auxiliary variable and the variable under study, we can expect that the ratio-type estimation strategies $(\hat{q}_{y,p,s}, P_{2,5}^{(+)}(s|w_1(q_{x,p}, d_x/2), w_2(q_{x,p}, d_x/4)))$ and $(\hat{q}_{y,p,s}, P_{2,5}^{(+)}(s|w_1(q_{x,p}, d_x/2)))$ are the most accurate among the considered ones. Let us note that the bias of the estimation may take quite high levels, even greater than 10% in the cases where p is close to 0 or to 1. Moreover, we can conclude that the estimation of the quantiles of the variable under study based on the quantile-

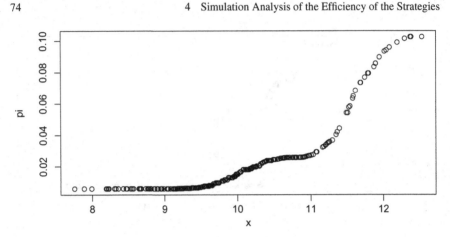

Fig. 4.16 The inclusion probabilities for $P_{2,5}^{(+)}(s|w_2(q_{x,0.9}, d_x/2))$, $q_{x,0.9} = 11.242$, $d_x = 0.932$, and $n = 6$. $X \sim N(10, 1)$. *Source* own preparation

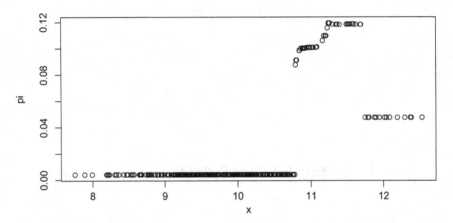

Fig. 4.17 The inclusion probabilities for $P_{2,5}^{(+)}(s|w_1(q_{x,0.9}, d_x/2), w_2(q_{x,0.9}, d_x/4))$, $q_{x,0.9} = 11.242$, $d_x = 0.932$, and $n = 6$. $X \sim N(10, 1)$. *Source* own preparation

dependent conditional sampling designs leads to more efficient estimation than the simple random sampling design or Sampford's (1967) design considered in Chap. 2.

Additionally, let us note that the sampling designs dependent on the sample quantile of the auxiliary variable can be generalized into the case where the auxiliary variable is multidimensional. Let $(x_{1,k}, x_{2,k})$, $k = 1, ..., N$, be observations of a two-dimensional auxiliary variable. We permit that values of the two-dimensional auxiliary are not necessarily positive, but they are highly correlated with a variable under study. Let $z_i = (x_{1,k} - \bar{x}_1)^2 + (x_{2,k} - \bar{x}_2)^2$, $k = 1, ..., N$, where $\bar{x}_i = \frac{1}{N}\sum_{k=1}^{N}$ is the population mean of the ith auxiliary variable, $i = 1, 2$. Let $Z_{(r)}$ be an order statistic from the simple random sample of the size n. Hence, the new sampling design proportional to values of the random variable $Z_{(r)}$ can be con-

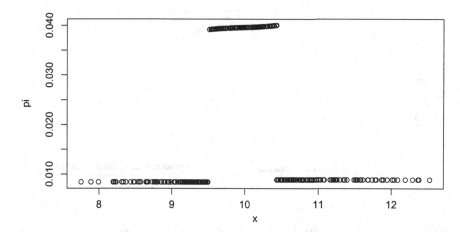

Fig. 4.18 The inclusion probabilities for $P_{2,5}^{(+)}(s|w_1(q_{x,0.5}, d_x/2))$, $q_{x,0.5} = 9.964$, $d_x = 0.932$, and $n = 6$. $X \sim N(10, 1)$. *Source* own preparation

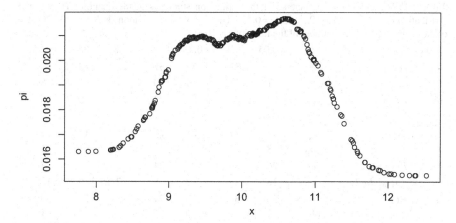

Fig. 4.19 The inclusion probabilities for $P_{2,5}^{(+)}(s|w_2(q_{x,0.5}, d_x/2))$, $q_{x,0.5} = 9.964$, $d_x = 0.932$, and $n = 6$. $X \sim N(10, 1)$. *Source* own preparation

structed as explained in Chap. 3. For instance, let the two-dimensional auxiliary variable be symmetrical around the point $(\bar{x}_1, \bar{x}_2)$. Hence, in this case, the sampling design $P(s|Z_r \leq c)$, $1 \leq r < n$, prefers sampling the population elements which are assigned to such observations of the auxiliary variables $(x_{1,k}, x_{2,k})$ that fulfil the inequality $z_i = (x_{1,k} - \bar{x}_1)^2 + (x_{2,k} - \bar{x}_2)^2 \leq c$. In particular, when we assume that $r = n - 1$, only one sample element is assigned to the auxiliary variables which are outside the circle with the origin in the point $\bar{x}_1, \bar{x}_2$ and the radius equal to $\sqrt{c}$ defined by the equation $z_i = (x_1 - \bar{x}_1)^2 + (x_2 - \bar{x}_2)^2 \leq c$. Hence, the defined sampling design prefers the population elements with observations of the auxiliary

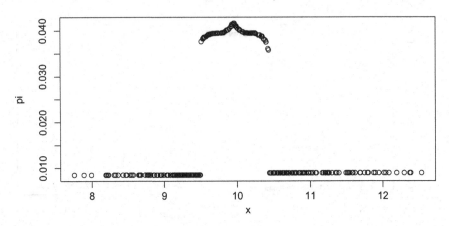

Fig. 4.20 The inclusion probabilities for $P_{2,5}^{(+)}(s|w_1(q_{x,0.5}, d_x/2), w_2(q_{x,0.5}, d_x/4))$, $q_{x,0.5} = 9.964$, $d_x = 0.932$, and $n = 6$. $X \sim N(10, 1)$. *Source* own preparation

Table 4.7 The efficiency coefficients of the estimation strategies of the quantile. The normal distribution $(X, Y) \sim N(10, 10, 1, 1, \rho = 0.95)$. The size of the populations is $N = 300$. The size of the sample $n = 6$. The number of the iterations is 1000

Strategies	p	0.05	0.1	0.2	0.5	0.8	0.9	0.95
$(\hat{q}_{y,p,s}, P_0(s))$	e_p	18	33	34	31	22	18	19
	b_p	1	1	14	0	4	1	5
$(q_{y,p,s}, P_{Sd}(s))$	e_p	93	88	108	74	73	92	76
	b_p	39	0	13	7	3	2	18
$(\hat{q}_{y,p,s}, P_{Sd}(s))$	e_p	15	27	23	20	15	15	13
	b_p	8	5	1	2	1	9	1
$(q_{y,p,s}, P(s|w_1))$,	e_p	26	36	27	34	39	33	27
$w_1 = w_1(q_{x,p}, d_x/2)$	b_p	18	8	2	8	7	9	12
$(q_{y,p,s}, P(s|w_2))$,	e_p	36	36	37	71	51	53	39
$w_2 = w_2(q_{x,p}, d_x/2)$	b_p	17	1	9	0	2	7	1
$(q_{y,p,s}, P(s|w_1, w_{2a}))$	e_p	20	19	28	36	28	16	14
$w_{2a} = w_2(q_{x,p}, d_x/4)$	b_p	20	0	1	4	1	3	8
$(\hat{q}_{y,p,s}, P(s|w_1))$	e_p	13	18	14	17	12	11	13
	b_p	17	5	13	1	0	1	14
$(\hat{q}_{y,p,s}, P(s|w_2))$	e_p	16	21	17	20	17	14	15
	b_p	0	3	7	1	1	9	15
$(\hat{q}_{y,p,s}, P(s|w_1, w_{2a}))$	e_p	14	15	14	18	11	8	10
	b_p	15	4	7	0	4	4	6

Source own calculations

variable from the inside of the just-defined circle. Of course, the circle equation can be replaced with the equation of an ellipse and so on. The evaluated result can be applied in space sampling.

4.6 Conclusions

The strategies considered are dependent on sample parameters of auxiliary variables. The strategies dependent on the sample moments (the sample mean and the sample variance) are more accurate than the simple random sample mean especially when the correlation coefficient between a variable under study and an auxiliary variable is high.

The sampling designs dependent on moments of multivariate auxiliary variables can be useful in the space sampling. Let us assume that coordinates of points identify the objects of a space population. In this case, the sampling design proportional to the sample generalized variance of the multidimensional auxiliary variable allows us to draw samples that are well-dispersed in the space population.

The accuracy of the unconditional strategies dependent on the order statistics (the quantile-type strategies) is usually not as high as the accuracy of the strategies dependent on the sample moments. In particular cases, the conditional versions of the quantile-type strategies can be superior to their unconditional versions, as shown in Fig. 4.5. Therefore, these conclusions suggest that looking for new constructions for the conditional strategies can improve the estimation accuracy.

In general, in the case of the estimation of the population mean, the relative efficiency of the unconditional strategies dependent on the sample parameters decreases when the sample size increases. This behaviour can be explained by the convergence of the sample parameters of an auxiliary variable to appropriate population parameters when the sample size and the population size increase infinitely.

The examples considered allow us to conclude that there is a possibility of modelling the relationship between the auxiliary variable and the inclusion probabilities. The shape of the relationship depends on the choice of the ranks of order statistics of the auxiliary variable as well as on the formulation of appropriate conditions that lead to drawing samples fulfilling assumed properties. In particular, there are application possibilities for the conditional version of the quantile-dependent sampling designs for inferences about domain means. Other formulations of the conditional sampling design can lead to drawing samples in such a way that the observed data fulfil some assumed formal properties, as was considered during the explanation of Figs. 4.1–4.9 and 4.11–4.20.

There is a possibility of generalizing the quantile sampling designs by case when they depend on at least four-order statistics, but this tends to lead to very complicated expressions with the inclusion probabilities. Moreover, there is significantly more computation time required to calculate those probabilities. The computer data processing time can be reduced in the case of appropriately constructed conditional versions of the sampling design.

The efficiency of the quantile-type estimation strategies considered is notably good in the case of the log-normal joint distribution of the auxiliary variable and the variable under study. Accordingly, larger-scale simulation analysis of the accuracy of the considered strategies is necessary on the basis of data generated as values of other skewed distributions, such as data used to model income or expenses. Moreover, such an analysis should lead to the formulation of rules of optimal choice, e.g. ranks of the order statistics of auxiliary variables.

Some aspects of the simulation analysis of the accuracy estimation can be generalized. In Chap. 3, we replaced the defined conditional sampling design proportional to the v_s under the assumption $v_s > c$ with conditional design proportional to the range $X_{(n)} - X_{(1)}$ under the condition that $X_{(n)} - X_{(1)} > c$. So, we can expect that this operation leads to improving accuracy of the regression estimator of the population mean. Let us note that the condition $X_{(n)} - X_{(1)} > c$ can be generalized into the following $X_{(u)} - X_{(r)} > c_1$ where $u > r$. In particular, we can expect that the regression estimator under the sampling design proportional to $X_{([n/2]+2)} - X_{([n/2])}$, provided that $X_{([n/2]+2)} - X_{([n/2])} > c_1$, can be more accurate than its unconditional version or the simple random sample mean, but it should be confirmed by simulation analysis.

In general, in the case of the estimation of the population mean, the relative efficiencies of the sampling strategies based on sampling designs dependent on sample moments of the auxiliary variable are usually not less than the efficiency of the strategies based on sampling design dependent on sample quantiles. However, some of the strategies based on conditional versions of quantile-dependent sampling designs are more efficient than the strategies based on moment-dependent sampling designs. In the case of the estimation of quantiles of the variable under study, the considered quantile-dependent conditional sampling designs significantly improve the accuracy of the estimation. Hence, we can suppose that the quantile-dependent sampling designs let us improve the accuracy of estimation of some special functions of quantiles of variables under study, such as Gini index or some poverty coefficients.

Our considerations and, especially, results of Chap. 4 lead us to expect that there are many possibilities for constructing new sampling designs dependent on sample moments or quantiles. They should improve estimation accuracy of some practically useful population parameters. From the other point of view, the practical application inspires us to looking for new sampling designs.

References

Arcos, A., Rueada, J. F., & Muñoz, J. F. (2007). An improved class of estimators of a finite population quantile in sample surveys. *Applied Mathematics Letters, 20*, 312–315.

Horvitz, D., & G., Thompson, D. J. (1952). A generalization of the sampling without replacement from finite universe. *Journal of the American Statistical Association, 47*, 663–685.

Rao, J. N. K., Kovar, J. G., & Mantel, H. J. (1990). On estimating distribution functions and quantiles from survey data using auxiliary information. *Biometrika, 77,* 365–375.

Sampford, M. R. (1967). On sampling without replacement with unequal probabilities of selection. *Biometrika, 54,* 499–513.

Särndal, C. E., Swensson, B., & Wretman, J. (1992). *Model Assisted Survey Sampling.* New York-Berlin-Heidelberg-London-Paris-Tokyo-Hong Kong- Barcelona-Budapest: Springer.

Wywiał, J. L. (2007). Simulation analysis of accuracy estimation of population mean on the basis of strategy dependent on sampling design proportionate to the order statistic of an auxiliary variable. *Statistics in Transition-New Series, 8*(1), 125–137.

Wywiał, J. L. (2016). *Contributions to Testing Statistical Hypotheses in Auditing.* Warsaw: PWN.

Chapter 5
Sampling Designs Dependent on a Continuous Auxiliary Variable

In contrast to the previous chapters, this chapter concerns the estimation of the continuous variable expected value of the variable under study, supported by a continuous auxiliary variable whose density function is known. Continuous versions of the definition of a sampling design and scheme are presented. In this context, the properties of the well-known Horvitz–Thompson statistic are considered. The two-dimensional gamma distribution of the variable under study and the auxiliary variable are taken into account. The mean of the variable under study is estimated by means of a ratio-type estimator. A continuous sampling design dependent on the order statistic of the uniformly distributed auxiliary variable is proposed, and its inclusion functions are derived. Finally, this sampling design is used to estimate the mean. The accuracy of this estimation is studied in the case of the continuous joint distribution of the variable under study and the auxiliary variable.

5.1 Introduction

The previous chapters discuss the estimation of population parameters based on a sampling design, which is defined as a function of the auxiliary variable values observed in the entire fixed and finite populations. A survey which utilizes this type of statistical inference is referred to as adopting the randomization approach. More general estimation procedures are based on a model approach where continuous variables are also considered (see, e.g. Thompson 1997).

The joint continuous distribution of the variable under study, as well as an auxiliary variable, was considered. Our goal was to estimate the mean of the variable being studied. To achieve this, a random sample was drawn according to a sampling design that was dependent on the observation of a continuous auxiliary variable in the entire population. The auxiliary variable values allowed for an estimation of the inclusion density function of the continuous version of the sampling design. The properties

J. L. Wywiał, *Sampling Designs Dependent on Sample Parameters of Auxiliary Variables*, SpringerBriefs in Statistics, https://doi.org/10.1007/978-3-662-63413-4_5

of the Horvitz–Thompson estimator under the proposed continuous sampling design were analyzed. The accuracy of the estimation strategy was analyzed using simulation experiments.

The continuous distribution function of the variable under study and the auxiliary variable (denoted by X and Y, respectively) were known or could be estimated. The values of X and Y were observed in the entire population of size N and in the sample, respectively. One example of an auxiliary variable observation in the entire population is given by tax data. Additional examples include area of cereals, electricity consumption in households, carbon dioxide emissions, or housing areas. Statistical inference based on an auxiliary variable could be applied in auditing. The joint continuous distribution of book values can be considered as an auxiliary variable observation, and the true values of the documents can be considered as observations of the variable under study. For example, Frost and Tamura (1986) and Wywiał (2018) considered the gamma distribution for modeling book values in statistical auditing.

Benhenni and Cambanis (1992) and Thompson (1997) considered continuous sampling for Monte Carlo integration. Moreover, continuous sampling designs were studied by Bąk (2014, 2018), Wilhelm et al. (2017), and Wywiał (2016). The accuracy of estimations based on stratified and systematic samples was studied by Cressie (1993) and Zubrzycki (1958). Sampling designs dependent on a positively valued continuous auxiliary variable, defined by Cox and Snell (1979), were applied to financial auditing. The basic definitions of continuous sampling designs and inclusion density functions were formulated by Cordy (1993), who also used the well-known Horvitz–Thompson (1952) statistic to estimate parameters. This chapter draws on these two extremely important sources.

5.2 Basic Definitions and Theorems

The main results regarding a continuous approach to sampling were obtained by Cordy (1993). The population is denoted by $U \subset R^q$, $q = 1, 2, \ldots$. In this chapter, only the case where $q = 1$ will be taken into account. The sample space, denoted by $S_n = U^n$, is the set of ordered samples denoted by $\mathbf{y} = [y_1, \ldots, y_n]$, $y_k \in U$, $k = 1, \ldots, n$, where y_i is the outcome of the variable observed in the first draw. $\mathbf{y}$ is a value of the n-dimensional random variable $\mathbf{Y} = [Y_1, \ldots, Y_n]$. The density function of the sampling design can be written as follows:

$$f(\mathbf{y}) = f(y_n, \ldots, y_i, y_{i-1}, \ldots, y_1) = f(y_1) \prod_{i=2}^{n} f(y_i | y_{i-1}, y_{i-2}, \ldots, y_1). \quad (5.1)$$

$f(y_i | y_{i-1}, y_{i-2}, \ldots, y_1), i = 1, \ldots, n - 1$ are the conditional density function of the selection y_i value in the ith draw (provided that the values $(y_{i-1}, y_{i-2}, \ldots, y_1)$ were drawn earlier).

Let $f_i(y)$ and $f_{i,j}(y, y')$, $y \in U$, $y' \in U$, be marginal density functions of $f(\mathbf{y})$, $j > i = 1, \ldots, n$. The inclusion functions of the first order and the second order are defined as follows, respectively:

$$\pi(y) = \sum_{i=1}^{n} f_i(y), \quad \pi(y, y') = \sum_{i=1}^{n} \sum_{j=1, j \neq i}^{n} f_{i,j}(y, y'), \quad y \in U, y' \in U \quad (5.2)$$

and $\int_U \pi(y) dy = n$, $\int_U \int_U \pi(y, y') dy dy' = n(n-1)$.

Let $g(y)$ be an integrable function $g : U \rightarrow R$. The estimated parameter is defined as follows:

$$\theta = \int_U g(y) dy. \quad (5.3)$$

The continuous version of the well-known Horvitz and Thompson (1952) estimator is

$$T_Y = \sum_{i=1}^{n} \frac{g(Y_i)}{\pi(Y_i)}. \quad (5.4)$$

Cordy (1993) proved the following theorems:

Theorem 5.1 *The statistic T_Y is an unbiased estimator for θ, if the function $g(y)$ is either bounded or non-negative, and $\pi(y) > 0$ for each $y \in U$.*

Theorem 5.2 *If the function $g(y)$ is bounded, $\pi(y) > 0$ for each $y \in U$, and $\int_U (1/\pi(y)) dy < \infty$, then*

$$\begin{aligned} V(T_Y) &= \int_U \frac{g^2(y)}{\pi(y)} dy + \int_U \int_U g(y) g(y') \frac{\pi(y, y') - \pi(y)\pi(y')}{\pi(y)\pi(y')} dy dy' \\ &= \int_U \frac{g^2(y)}{\pi(y)} dy + \int_U \int_U g(y) g(y') \frac{\pi(y, y')}{\pi(y)\pi(y')} dy dy' - \theta^2. \end{aligned} \quad (5.5)$$

When $\pi(y_i, y_j) > 0$ for all $y_i, y_j \in U$, $i \neq j = 1, \ldots, n$, the unbiased estimator of the variance is

$$\hat{V}(T_Y) = \sum_{i=1}^{n} \frac{g^2(Y_i)}{\pi^2(Y_i)} + \sum_{i=1}^{n} \sum_{j=1, i \neq j}^{n} g(Y_i) g(Y_j) \frac{\pi(Y_i, Y_j) - \pi(Y_i)\pi(Y_j)}{\pi(Y_i, Y_j)\pi(Y_i)\pi(Y_j)}.$$

In particular, when $h(y)$ is a density function and $g(y) = \eta(y)h(y)$, then $\theta = E(\eta(Y))$. If $\eta(y) = y$, then $\theta = E(Y)$.

If $Y_1, \ldots, Y_n$ is a random sample from a distribution with density $f(\mathbf{y})$, then

$$f(\mathbf{y}) = \prod_{i=1}^{n} f(y_i), \quad \pi(y) = nf(y), \quad \pi(y, y') = n(n-1)f(y)f(y'), \qquad (5.6)$$

$$T_{\mathbf{Y}} = \frac{1}{n} \sum_{i=1}^{n} \frac{\eta(Y_i)h(Y_i)}{f(Y_i)}, \quad E(T_{\mathbf{Y}}) = \theta, \qquad (5.7)$$

$$\begin{aligned}
V(T_{\mathbf{Y}}) &= \frac{1}{n} \left(\int_U \frac{\eta^2(y)h^2(y)}{f(y)} dy - \theta^2 \right) \\
&= \frac{1}{n} \left(E\left(\frac{\eta^2(Y)h^2(Y)}{f^2(Y)} \right) - E^2\left(\frac{\eta(Y)h(Y)}{f(Y)} \right) \right) \\
&= \frac{1}{n} V\left(\frac{\eta(Y)h(Y)}{f(Y)} \right).
\end{aligned} \qquad (5.8)$$

Sampling design $f(y_n, \ldots, y_1)$, given by (5.6), provides what is known as the *importance sample* utilized by Bucklew (2004) and Ripley (1987). When the importance sample is drawn from the density $h(y)$ and $\eta(y) = y$, then it becomes the well-known simple random sample, defined as the sequence of independent and identically distributed random variables, and $\theta = E(Y) = \mu_y$ is estimated using the following statistic:

$$T_Y = \bar{Y} = \frac{1}{n} \sum_{i=1}^{n} Y_i, \quad V(T_Y) = V(\bar{Y}) = \frac{1}{n} V(Y) \qquad (5.9)$$

where $V(Y) = \int_{-\infty}^{\infty} (y - E(Y))^2 f(y) dy$.

Let $h(x, y), (x, y) \in U \subseteq R^2$, be the density function. The marginal densities and the conditional density will be denoted by $h_1(x), h_2(y)$, and $h(y|x) = h(x, y)/h_1(x)$. Moreover, let $\mu_y = E(Y) = \int_{-\infty}^{-\infty} yh_2(y)dy$, $\mu_x = E(X) = \int_{-\infty}^{-\infty} xh_1(x)dx$, $E(Y|x) = \int_{-\infty}^{-\infty} yh(y|x)dy$, and $V(Y|x) = \int_{-\infty}^{-\infty} (y - E(Y|x))^2 h(y|x)dy$. Wywiał (2020) considered estimation of parameter $\theta = \int_{-\infty}^{\infty} g(x)dx$ (see, expression (5.3), where

$$g(x) = E(\eta(Y)|x)h_1(x) = h_1(x) \int_{-\infty}^{\infty} \eta(y)h(y|x)dy.$$

When we set $\eta(y) = y$, then

$$\theta = \mu_y = \int_{-\infty}^{\infty} E(Y|x)h_1(x)dx = \int_{-\infty}^{\infty} \int_{-\infty}^{\infty} yh(y|x)h_1(x)dxdy. \tag{5.10}$$

Parameter μ_y is estimated using the following statistic:

$$T_{\mathbf{X},\mathbf{Y}} = \sum_{i=1}^{n} \frac{Y_i h_1(X_i)}{\pi(X_i)} \tag{5.11}$$

where $\{X_i, i = 1, \dots, n\}$ is the sample drawn according to the sampling design defined by expression (5.1) where y_i should be replaced by x_i. Let us assume that

$$h(y|x) = h(y_1, \dots, y_n|x_1, \dots, x_n) = \prod_{i=1}^{n} h(y_i|x_i). \tag{5.12}$$

Let $E_{h(Y/X)}(T_{\mathbf{X},\mathbf{Y}}) = \int_{-\infty}^{\infty} t_{x.y} f(y|x)dy$, $E_{f(X)}(T_{\mathbf{X},\mathbf{Y}}) = \int_{-\infty}^{\infty} t_{x.y} f(x)dx$, and $V(T_{\mathbf{X},\mathbf{Y}}) = E_{f(X)} E_{h(Y/X)}(T_{\mathbf{X},\mathbf{Y}} - E_{h(Y/X)}(T_{\mathbf{X},\mathbf{Y}}))^2$. Using Cordy's proofs of Theorems 5.1 and 5.2 Wywiał (2020) proved the following:

Theorem 5.3 *If $E(Y) < \infty$ and $\pi(x) > 0$ for all $(x, y) \in U$ and assumption (5.12) holds, then $E_{f(X)} E_{h(Y/X)}(T_{\mathbf{X},\mathbf{Y}}) = \mu_y$.*

Theorem 5.4 *If the function $E(Y)$ is bounded, $\pi(y) > 0$ for each $(x, y) \in U$, and $\int_U (1/\pi(y))dy < \infty$, then*

$$V(T_{\mathbf{X},\mathbf{Y}}) = \int_U \frac{V(Y|x)h_1^2(x)}{\pi(x)}dx + \int_U \frac{E^2(Y|x)h_1^2(x)}{\pi(x)}dx + A \tag{5.13}$$

where

$$A = \int_U \int_U E(Y|x)h_1(x)E(Y|x')h_1(x')\frac{\pi(x, x') - \pi(x)\pi(x')}{\pi(x)\pi(x')}dxdx'$$

or

$$A = \int_U \int_U E(Y|x)h_1(x)E(Y|x')h_1(x')\frac{\pi(x, x')}{\pi(x)\pi(x')}dxdx' - E^2(Y).$$

5.3 The Inclusion Function Proportional to the Values of the Auxiliary Variable

Cox and Snell (1979) considered the following sampling design:

$$f(x_1, \ldots, x_n) = \prod_{i=1}^{n} f(x_i), \qquad f(x_i) = \frac{x_i h_1(x_i)}{\mu_x} \qquad (5.14)$$

where $\mu_x = E(X) = E(X_i)$ for all $i = 1, \ldots, n$. In this case according to (5.6), where y should be replaced by x, the inclusion function is proportional to the value of the auxiliary variable because $\pi(x) = \frac{n x h_1(x)}{\mu_x}$. The expression (5.11), Theorems 5.3 and 5.4 lead to the following:

$$T_{\mathbf{X},\mathbf{Y}} = \hat{Y}_R = \frac{\mu_x}{n} \sum_{i=1}^{n} \frac{Y_i}{X_i}, \qquad E(\hat{Y}_R) = \mu_y, \qquad (5.15)$$

$$V(T_{\mathbf{X},\mathbf{Y}}) = \frac{1}{n} \left(\mu_x \int_U \frac{V(Y|x) h_1(x)}{x} dx + \mu_x \int_U \frac{E^2(Y|x) h_1(x)}{x} dx - \mu_y^2 \right)$$

$$= \frac{\mu_x}{n} \int_U \frac{V(Y|x) h_1(x)}{x} dx + \frac{\mu_x}{n} V\left(\frac{E(Y|x)}{x} \right). \qquad (5.16)$$

Hence, statistic $\hat{Y}_R$ is an unbiased ratio-type estimator of μ_y.

When parameter μ_x and other parameters of the auxiliary variable density function are known, the sample can be selected. Wywiał (2020) considered the case where the parameters of the auxiliary variable density function were estimated. Moreover, the author proposed to assess the auxiliary variable density using the well-known kernel estimator.

In general, in practical research, it is assumed that $\mathbf{x} = [x_1, \ldots, x_k, \ldots, x_N]$ are known values. The function $h_1(x)$ is known or estimated. The sample $\mathbf{x}_s = [x_1, \ldots, x_k, \ldots, x_n]$ is selected as the sub-vector of $\mathbf{x}$ according to the sampling design defined by expression (5.14). To achieve this, values of vector $\mathbf{x}' = [x_1', \ldots, x_n']$ are generated using the quantile functions $x' = F^{-1}(u)$, where u is the value of the uniformly distributed variable at interval $[0; 1]$, $F(x) = \int_{-\infty}^{x} f(t) dt$. The elements of $\mathbf{x}_s$ are selected from $\mathbf{x}$ according to

$$x_k = arg \min_{j=1,\ldots,N} |x_j - x_k'|. \qquad (5.17)$$

This algorithm could yield a sample with duplicate elements. To avoid these duplicates, the sample should be rejected and the algorithm repeated until a sample with no duplicates is obtained.

An additional algorithm, which leads to the drawing of $\mathbf{x}_s$ without repetition, is explained by expression:

$$\mathbf{x}_s = arg \min_{\mathbf{x}_s \in \mathbf{X}_s} (\mathbf{x}_s - \mathbf{x}'_s)(\mathbf{x}_s - \mathbf{x}'_s)^T \qquad (5.18)$$

where $\mathbf{x}_s$ consists of all n-element combinations selected without replacement from $\mathbf{x}$. The complete data $\mathbf{d} = [(x_1, y_1), \ldots, (x_n, y_n)]$ are evaluated after observing values $y_j, j = 1, \ldots, n$, which are attached to the appropriate elements of vector $\mathbf{x}_s$.

Wywiał (2020) analyzed the example of estimation in the case of McKay's (1934) bivariate gamma distribution. This distribution has the following density function (see also Ghirtis 1967; Kotz et al. 2000):

$$h(x, y) = \frac{c^\theta}{\Gamma(\theta_y)\Gamma(\theta_x)} y^{\theta_y - 1}(x - y)^{\theta_x - 1} e^{-cx}, \quad x > y > 0 \qquad (5.19)$$

where $\theta = \theta_x + \theta_y$. The auxiliary variable marginal distribution shows the following gamma density function:

$$f(x) = \frac{c^{\theta_x}}{\Gamma(\theta_x)} x^{\theta_x - 1} e^{-cx}, \qquad E(X) = \mu_x = \frac{\theta_x}{c}.$$

According to expression (5.14), the sampling design density function is as follows:

$$f(x) = \frac{c^{\theta_x + 1}}{\Gamma(\theta_x + 1)} x^{\theta_x} e^{-cx}. \qquad (5.20)$$

$f(x)$ is also a density function of the gamma distribution, with shape and scale parameters equal to $\theta_x + 1$ and c, respectively. Parameter $\mu_y = E(Y) = \frac{\theta_y}{c}$ is the goal of the estimation. The ratio estimator, defined by (5.15), has the following variance (derived based on Theorem 5.4):

$$V(\hat{Y}_R) = \frac{\theta_y \theta_x}{nc^2(\theta + 1)}. \qquad (5.21)$$

The estimator $\hat{Y}_R$ is more precise, than given by (5.9) the simple random sample mean, which is denoted by $\bar{Y}$ because the relative efficiency coefficient takes the following form:

$$deff(\hat{Y}_R) = 100\% \frac{V(\hat{Y}_R)}{V(\bar{Y})} = \frac{100\% \theta}{\theta_x + 1} < 100\%. \qquad (5.22)$$

The variance $V(\hat{Y}_R)$ could be estimated using the bootstrap method or by replacing the parameters with their appropriate estimators in expression (5.21).

Under the assumed McKay distribution, Wywiał (2020) considered the accuracy of the two variants of the estimator given by (5.15), where μ_x was replaced by $\bar{x} = \frac{1}{N}\sum_{i=1}^{N} x_i$. These variants were dependent on the sampling design used. The first variant was based on expression (5.20) where the parameters were estimated based on the vector $\mathbf{x}$. The second variant was dependent on the kernel estimator of the density function $h_1(x)$. For the several parameters of the McKay distribution, the samples were selected. This allowed for the variance of the estimators to be assessed. The results of the simulation analysis showed that both estimators have comparable accuracy and that they are superior to the simple random sample mean. The estimation that was based on the sampling design which was dependent on the kernel estimator was recommended because it did not depend on the auxiliary variable density function.

5.4   Sampling Designs as a Function of an Order Statistic

In Chap. 3, the discrete sampling designs that were dependent on the order statistics of all values of a positive auxiliary variable that was observed in an entire fixed and finite populations were considered. The sampling designs were deemed to be useful in practice because the results of the simulation analysis developed in Chap. 4 confirmed the accuracy of the estimation based on these designs. A continuous version of one of these sampling designs will be considered in this section.

5.4.1   General Results

Let $h(x)$, $x \in R_+$ be the density function of a continuous random variable X. The order statistic of rank r, determined based on a simple random sample of size n, drawn from X, will be denoted by $X_{r:n}$ and its density function by $h_{r:n}(x)$.

The following truncated distributions of X in point $X_{r:n} = x_r$ are considered:

$$f_1(x|x_r) = \frac{h(x)}{H(x_r)} \quad \text{for} \ \ 0 \le x \le x_r \ \ \text{and} \ \ f_1(x|x_r) = 0 \ \ \text{for} \ \ x > x_r, \quad (5.23)$$

$$f_2(x|x_r) = \frac{h(x)}{1 - H(x_r)} \quad \text{for} \ \ x > x_r \ \ \text{and} \ \ f_2(x|x_r) = 0 \ \ \text{for} \ \ 0 \le x \le x_r$$
$$(5.24)$$

where $H(x_r) = P(X < x_r) = \int_0^{x_r} h(x)dx$. Hence, $f_1(x|x_r)$ and $f_2(x|x_r)$ are the densities of the right and left truncated distribution of X at point x_r, respectively.

In Sect. 3.2, the sampling scheme for the selection of the sample according to the discrete version of the quantile sampling design was shown. The continuous version of this sampling scheme is as follows. Firstly, we consider the following density function:

$$f_{*r}(x) = \frac{x h_{r:n}(x)}{E(X_{r:n})} \tag{5.25}$$

of the random variable denoted by X_r. This function is the particular case of expression (5.14). The value x_r, generated according to the density function $f_{*r}(x)$ of X_r, is the first observation of sample $X = [X_1, \ldots, X_{r-1}, X_r, X_{r+1}, \ldots, X_n]$. Next, if $1 < r \le n$, the values $[x_1, \ldots, x_{r-1}]$ of variables $[X_1, \ldots, X_{r-1}]$ are drawn independently from the density $f_1(x|x_r)$. Similarly, when $1 \le r < n$, the values $[x_{r+1}, \ldots, x_n]$ of variables $[X_{r+1}, \ldots, X_n]$ are drawn independently from the density $f_2(x|x_r)$. Hence, the sampling scheme is defined by the density functions: $f_1(x|x_r)$, $f_{*r}(x)$, and $f_2(x|x_r)$. Let $x \in D = R_+^n + \{0\}$. The following sampling design will be considered:

$$f(x) = f(x_1, \ldots, x_{r-1}, x_r, x_{r+1}, \ldots, x_n)$$
$$= \begin{cases} f_{*r}(x_r) \prod_{i=1}^{r-1} f_1(x_i|x_r) \prod_{i=r+1}^{n} f_2(x_i|x_r) & \text{for } x \in D, \\ 0 & \text{for } x \notin D. \end{cases} \tag{5.26}$$

This function will be named the quantile sampling design.

Theorem 5.5

$$\pi(x) = (r-1)f_1(x) + f_{*r}(x) + (n-r)f_n(x) \tag{5.27}$$

where $f_{*r(x)}$ is given by (5.25) and

$$f_1(x) = f_i(x) = \int_x^\infty f_1(x|x_r) f_{*r}(x_r) dx_r, \quad i = 1, \ldots, r-1, r > 1, \tag{5.28}$$

$$f_n(x) = f_i(x) = \int_0^x f_2(x|x_r) f_{*r}(x_r) dx_r, \quad i = r+1, \ldots, n, r < n. \tag{5.29}$$

Proof Expressions (5.23)–(5.26) let us derive the marginal density $f_i(x)$ as follows. If $i < r$, then $x_i < x_r$ and

$$f_i(x_i)$$

$$= \left(\prod_{k=1,k\neq i}^{r-1} \int_0^\infty f_1(x_k|x_r) dx_k \right) \int_0^\infty f_1(x_i|x_r) f_{*r}(x_r) dx_r \left(\prod_{k=r+1}^{n} \int_0^\infty f_2(x_k|x_r) dx_k \right)$$

$$= \left(\prod_{k=1,k\neq i}^{r-1} \int_0^{x_r} f_1(x_k|x_r) dx_k \right) \int_0^\infty f_1(x_i|x_r) f_{*r}(x_r) dx_r \left(\prod_{k=r+1}^{n} \int_{x_r}^\infty f_2(x_k|x_r) dx_k \right)$$

$$= \int_0^\infty f_1(x_i|x_r)f_{*r}(x_r)dx_r = \int_{x_i}^\infty f_1(x_i|x_r)f_{*r}(x_r)dx_r \quad \text{for} \quad i = 1, \ldots, r-1$$

(5.30)

because $\int_0^{x_r} f_1(x_k|x_r)dx_k = 1$ for $i = 1, \ldots, r-1$ and $\int_{x_r}^\infty f_1(x_k|x_r)dx_k = 1$ for $i = r+1, \ldots, n$. Next

$$f_i(x_i) = \int_0^\infty f_1(x_i|x_r)f_{*r}(x_r)dx_r = \int_{x_i}^\infty f_1(x_i|x_r)f_{*r}(x_r)dx_r \quad \text{for} \quad i = 1, \ldots, r-1$$

because $f_1(x_i|x_r)f_{*r}(x_r) > 0$ for $(x_i, x_r) \in D_1 = \{(x, y) : 0 \le x < \infty, x \le y < \infty\}$ and $f_1(x_i, x_r) = 0$ for $(x_i, x_r) \notin D_1$. This leads to (5.28).

If $i > r$, then $x_i > x_r$ and

$$f_i(x_i)$$

$$= \left(\prod_{k=1}^{r-1}\int_0^{x_r} f_1(x_k|x_r)dx_k\right)\int_0^\infty f_2(x_i|x_r)f_{*r}(x_r)dx_r \left(\prod_{k=r+1,k\neq i}^n\int_{x_r}^\infty f_1(x_k|x_r)dx_k\right)$$

$$= \int_0^\infty f_2(x_i|x_r)f_{*r}(x_r)dx_r = \int_0^{x_r} f_2(x_i|x_r)f_{*r}(x_r)dx_r \quad \text{for} \quad i = 1, \ldots, r-1$$

(5.31)

because $f_2(x_i, x_r) = f_2(x_i|x_r)f_{*r}(x_r) > 0$ for $(x_i, x_r) \in D_2 = \{(x, y) : 0 \le x < \infty, 0 \le y < x\}$ and $f_2(x_i, x_r) = 0$ for $(x_i, x_r) \notin D_2$. This leads to expression (5.29).

Moreover, $f_i(x) = f_j(x)$ for $i = 1, \ldots, r-1$, $i \neq j$, $j = 1, \ldots, r-1$, and $f_i(x) = f_j(x)$ for $i = r+1, \ldots, n$, $i \neq j$, $j = r+1, \ldots, n$. Finally, expressions (5.1), where y should be replaced by x, lead to (5.27). This completes the proof.

Theorem 5.6

$$\pi(x, y) = (r-1)(r-2)f_{1,2}(x, y) + (n-r)(n-r-1)f_{n-1,n}(x, y)$$
$$+ 2(r-1)(n-r)f_{r-1,r+1}(x, y) + 2(r-1)f_{r-1,r}(x, y)$$
$$+ 2(n-r)f_{r,r+1}(x, y)$$

(5.32)

where, if $y \ge x$:

$$f_{r-1,r}(x, y) = f_{i,r}(x, y) = f_1(x|y)f_{*r}(y), \quad for\ i < r,$$

(5.33)

$$f_{r,r+1}(x, y) = f_{r,i}(x, y) = f_2(x|y)f_{*r}(y), \quad for\ i > r,$$

(5.34)

$$f_{1,2}(x, y) = f_{i,j}(x, y) = 2 \int_y^\infty f_1(x|x_r) f_1(y|x_r) f_{*r}(x_r) dx_r \qquad (5.35)$$

for $2 < r \le n$ and $x < x_r$, $y < x_r$, $i = 1, \ldots, r-1$, $j = 1, \ldots, r-1$ and $i \ne j$,

$$f_{n-1,n}(x, y) = f_{i,j}(x, y) = 2 \int_0^x f_2(x|x_r) f_2(y|x_r) f_{*r}(x_r) dx_r \qquad (5.36)$$

for $1 \le r < n-1$ and $x > x_r$, $y > x_r$, $i = r+1, \ldots, n$, $j = r+1, \ldots, n$ and $i \ne j$,

$$f_{r-1,r+1}(x, y) = f_{i,j}(x, y) = \int_x^y f_1(x|x_r) f_2(y|x_r) f_{*r}(x_r) dx_r \qquad (5.37)$$

for $1 < r < n$, $x < x_r < y$, $i = 1, \ldots, r-1$, $j = r+1, \ldots, n$, and $f_{i,j}(x, y) = f_{j,i}(x, y)$ for $i = 1, \ldots, n$, $j = 1, \ldots, n$, $i \ne j$.

Proof The derivation of expressions (5.33) and (5.34) is similar to that given by (5.30) and (5.31). Random variables $X = X_i$ and $Y = X_j$ are independent and have the same density functions $f_1(x|x_r)$ and $f_1(y|x_r)$, respectively. The well-known properties of order statistics (see, e.g. David and Nagaraja 2003) let us write that $2! f_1(x|x_r) f_1(y|x_r)$ is the density function of order statistics. This lead to the following inclusion function:

$$f_{i,j}(x, y) = \int_0^\infty f_1(x|x_r) f_1(y|x_r) f_{*r}(x_r) dx_r$$

$$= I(y - x) \int_y^\infty 2! f_1(x|x_r) f_1(y|x_r) f_{*r}(x_r) dx_r$$

$$+ I(x - y) \int_x^\infty 2! f_1(x|x_r) f_1(y|x_r) f_{*r}(x_r) dx_r$$

$$= 2 \int_{yI(y-x)+x(1-I(y-x))}^\infty f_1(x|x_r) f_1(y|x_r) f_{*r}(x_r) dx_r$$

where

$$I(u) = \begin{cases} 1 & \text{for } u \ge 0 \\ 0 & \text{for } u < 0 \end{cases}$$

where $2 < r \leq n$ and $x < x_r$, $y < x_r$, $i = 1, \ldots, r-1$, $j = 1, \ldots, r-1$ and $i \neq j$. This leads to a simplified version of the expression given by (5.35) under the assumption: $y > x$.

Similar to the above expression, we derive the following:

$$f_{i,j}(x, y) = \int_0^\infty f_2(x|x_r) f_2(y|x_r) f_{*r}(x_r) dx_r$$

$$= 2I(y - x) \int_0^x f_2(x|x_r) f_2(y|x_r) f_{*r}(x_r) dx_r$$

$$+ 2I(x - y) \int_0^y f_2(x|x_r) f_2(y|x_r) f_{*r}(x_r) dx_r$$

$$= 2 \int_0^{xI(y-x)+(1-I(y-x))} f_2(x|x_r) f_2(y|x_r) f_{*r}(x_r) dx_r$$

for $1 \leq r < n - 1$ and $x > x_r$, $y > x_r$, $i = r+1, \ldots, n$, $j = r+1, \ldots, n$ and $i \neq j$. A simpler version of this result is given by (5.36) for $y > x$.

When $1 < r < n$, $x < x_r < y$ or $y < x_r < x$ $i = 1, \ldots, r-1$, $j = r+1, \ldots, n$, then

$$f_{i,j}(x, y) = +I(y - x) \int_x^y f_1(x|x_r) f_2(y|x_r) f_{*r}(x_r) dx_r$$

$$+ I(x - y) \int_y^x f_1(x|x_r) f_2(y|x_r) f_{*r}(x_r) dx_r$$

$$= \int_{xI(y-x)+(1-I(y-x))}^{yI(y-x)+x(1-I(y-x))} f_1(x|x_r) f_2(y|x_r) f_{*r}(x_r) dx_r.$$

This leads to (5.37).

5.4.2 Uniform Distribution

Let us consider the following density function of uniform distribution $h(x) = 1$ for $x \in [0; 1]$ and $h(x) = 0$ for $x \notin [0; 1]$. In this case, the distribution of the rth order statistic from the simple sample of size n drawn from $h(x)$ is (see, e.g. David and Nagaraja 2003)

$$h_{r:n}(x) = \begin{cases} \frac{n!}{(r-1)!(n-r)!}x^{r-1}(1-x)^{n-r} & \text{for } x \in [0; 1], \\ 0 & \text{for } x \notin [0; 1] \end{cases} \tag{5.38}$$

with the expected value of $E(X_{r:n}) = \frac{r}{n+1}$. This and expression (5.25) lead to the following density:

$$f_{*r}(x) = \begin{cases} (n+1)\binom{n}{r}x^r(1-x)^{n-r} & \text{for } x \in [0; 1], \\ 0 & \text{for } x \notin [0; 1]. \end{cases} \tag{5.39}$$

Hence, the first value observed in the sample is drawn according to density $f_{*r}(x) = h_{(r+1):(n+1)}(x)$, and it will be denoted by x_r. The subsequent values $x_i, i = 1, \ldots, r - 1, r + 1, \ldots, n$ observed in the sample are generated according to the densities:

$$f_1(x|x_r) = \begin{cases} \frac{1}{x_r} & \text{for } x \in [0; x_r], \\ 0 & \text{for } x \notin [0; x_r] \end{cases} \tag{5.40}$$

provided that $r > 1$ and $i = 1, \ldots, r - 1$. Similarly, we evaluate

$$f_2(x|x_r) = \begin{cases} \frac{1}{1-x_r} & \text{for } x \in [x_r; 1], \\ 0 & \text{for } x \notin [x_r; 1] \end{cases} \tag{5.41}$$

provided that $r < n$ and $i = r + 1, \ldots, n$.

Let $D_0 = \{x = (x_1, \ldots, x_n) : 0 \le x_i < x_r < x_j \le 1, i = 1, \ldots, r - 1, j = r + 1, \ldots, n\}$. According to expression (5.26), we derive the following sampling design:

$$f(x_1, \ldots, x_n) = \begin{cases} (n+1)\binom{n}{r}x_r & \text{for } x \in D_0, \\ 0 & \text{for } x \notin D_0. \end{cases} \tag{5.42}$$

Theorem 5.7 *The density functions of the marginal probability distribution for the sampling design given by (5.42) are as follows: If $i = r$ then $f_i(x) = f_{*r(x)}$, given by (5.39), and if $1 < r \le n$ and $i = 1, \ldots, r - 1$ then*

$$f_i(x) = f_1(x) = (n+1)\binom{n}{r}\sum_{k=0}^{n-r}(-1)^k\binom{n-r}{k}\frac{1-x^{k+r}}{k+r}, \tag{5.43}$$

if $1 \le r < n$ and $i = r + 1, \ldots, n$:

$$f_i(x) = f_n(x) = (n+1)\binom{n}{r}\sum_{k=0}^{n-r-1}(-1)^k\binom{n-r-1}{k}\frac{x^{k+r+1}}{k+r+1}. \tag{5.44}$$

If $x < y$,

$$f_{i,j}(x, y) = f_{n-1,n}(x, y)$$

$$= 2(n+1)\binom{n}{r}\sum_{k=0}^{n-r-2}(-1)^k\binom{n-r-2}{k}\frac{x^{k+r+1}}{k+r+1} \qquad (5.45)$$

for $1 \leq r < n - 1$, $x_r < x$, $x_r < y$, $i = r + 1, \ldots, n$, $j = r + 1, \ldots, n$, and $i \neq j$;

$$f_{i,j}(x, y) = f_{1,2}(x, y) = 2(n+1)\binom{n}{r}\sum_{k=0}^{n-r}(-1)^k\binom{n-r}{k}\frac{1 - y^{k+r-1}}{k+r-1} \qquad (5.46)$$

for $1 < r \leq n$, $x < x_r$, $y < x_r$, $i = 1, \ldots, r-1$, $j = 1, \ldots, r-1$, $i < r$, $j < r$, and $i \neq j$;

$$f_{i,j}(x, y) = f_{r-1,r+1}(x, y)$$

$$= (n+1)\binom{n}{r}\sum_{k=0}^{n-r-1}(-1)^k\binom{n-r-1}{k}\frac{y^{k+r} - x^{k+r}}{k+r} \qquad (5.47)$$

for $0 < r < n$, $x < x_r < y$, and $i = 1, \ldots, r-1$, $j = r + 1, \ldots, n$;

$$f_{i,r}(x, y) = f_{r-1,r}(x, y) = (n+1)\binom{n}{r}y^{r-1}(1-y)^{(n-r)}, \qquad (5.48)$$

for $1 < r \leq n$, $y = x_r$, and $i = 1, \ldots, r-1$;

$$f_{r,j}(x, y) = f_{r,r+1}(x, y) = (n+1)\binom{n}{r}x^r(1-x)^{(n-r-1)} \qquad (5.49)$$

for $1 \leq r < n$, $x_r = x$, and $j = r + 1, \ldots, n$.

Proof Expressions (5.39) and (5.40) let us state that $f_1(x|x_r)f_{*r}(x_r)$ is positive for $x \leq x_r \leq 1$ and $0 \leq x \leq 1$. This and Theorem 5.5 lead to the following derivation. For $1 < r \leq n$ and $i = 1, \ldots, r-1$, we have

$$f_1(x) = \int_x^1 f_1(x|x_r)f_{*r}(x_r)dx_r = \frac{(n+1)!}{r!(n-r)!}\int_x^1 x_r^{r-1}(1-x_r)^{n-r}dx_r$$

$$= \frac{(n+1)!}{r!(n-r)!}\sum_{k=0}^{n-r}(-1)^k\binom{n-r}{k}\int_x^1 x_r^{k+r-1}dx_r.$$

This leads to expression (5.43).

Similarly, $f_2(x|x_r)f_{*r}(x_r)$ is positive for $0 \leq x_r < x$ and $0 \leq x \leq 1$. Let $1 \leq r < n$ and $i = r + 1, \ldots, n$. These assumptions and Theorem 5.5 lead to the following derivation:

$$f_n(x) = \int_0^x f_2(x|x_r)f_{*r}(x_r)dx_r = \frac{(n+1)!}{r!(n-r)!}\int_0^x x_r^r(1-x_r)^{n-r-1}dx_r$$

$$= (n+1)\binom{n}{r}\sum_{k=0}^{n-r-1}(-1)^k\binom{n-r-1}{k}\int_0^x x_r^{k+r}dx_r.$$

This leads to (5.44).

Theorem 5.6 and expressions (5.39)–(5.41) and (5.36) lead to the following: for $1 \le r < n-1, x_r < x, x_r < y, x < y, i = r+1, \ldots, n, j = r+1, \ldots, n$, and $i \ne j$, we have

$$f_{i,j}(x, y) = 2(n+1)\binom{n}{r}\int_0^x x_r^r(1-x_r)^{n-r-2}dx_r$$

$$= 2(n+1)\binom{n}{r}\sum_{k=0}^{n-r-2}(-1)^k\binom{n-r-2}{k}\int_0^x x_r^{k+r}dx_r.$$

This leads to expression (5.45).

Similar to the above derivation, we evaluate expressions (5.46)–(5.49).

Figures 5.1, 5.2, 5.3, 5.4, 5.5, and 5.6 show examples of the inclusion function shapes for several sample sizes n and ranks r of the order statistic. An analysis of these figures leads to the following conclusions. All inclusion functions are increasing functions of argument x. Under fixed n, the inflection point of the inclusion function moves to the right when the rank of the order statistic increases. Under fixed p (the rank of the quantile), when the sample size n increases, the inclusion function takes small values on the longer segment of the argument shifting from zero toward one.

Corollary 1 *When $r = 1$ and $n = 2$, Theorem 5.7 and expressions (5.39), (5.42), (5.43), and (5.49) let us write*

$$\begin{cases} f(x, y) = 6x, & \text{for } 0 < x < y < 1, \\ f_1(x) = 6x(1-x), & f_2(x) = 3x^2, \quad \pi(x) = 3x(2-x), \\ \pi(x, y) = 2f_{1,2}(x, y) = 12x & \text{for } x < y. \end{cases}$$

When $r = n = 2$, expressions (5.39), (5.42), (5.43), and (5.48) lead to

$$\begin{cases} f(x, y) = 3y, & \text{for } 0 < x < y < 1, \\ f_1(x) = \frac{3}{2}(1-x^2), & f_2(x) = 3x^2, \quad \pi(x) = \frac{3}{2}(1+x^2), \\ \pi(x, y) = 2f_{1,2}(x, y) = 6y & \text{for } x < y. \end{cases}$$

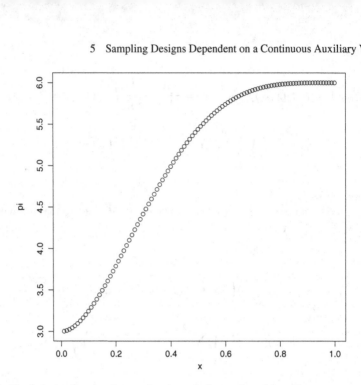

Fig. 5.1 The inclusion function for $n = 5$ and $r = 2$. *Source* Own preparation

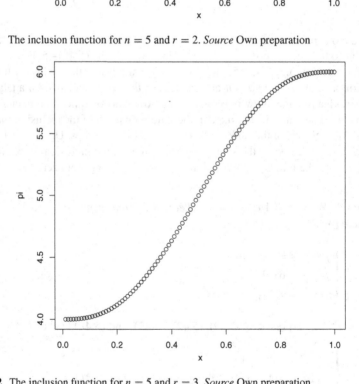

Fig. 5.2 The inclusion function for $n = 5$ and $r = 3$. *Source* Own preparation

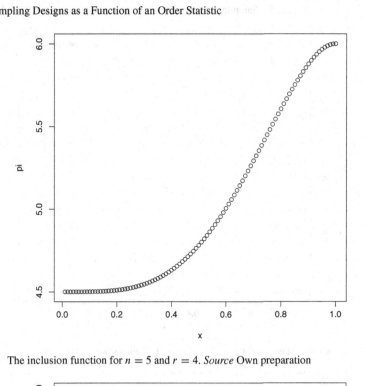

Fig. 5.3 The inclusion function for $n = 5$ and $r = 4$. *Source* Own preparation

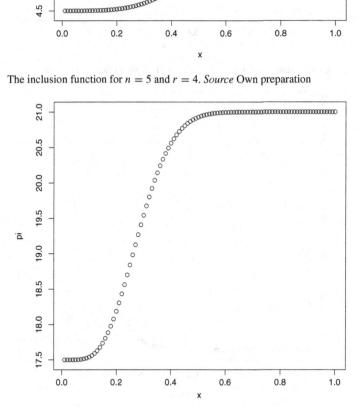

Fig. 5.4 The inclusion function for $n = 20$ and $r = 6$. *Source* Own preparation

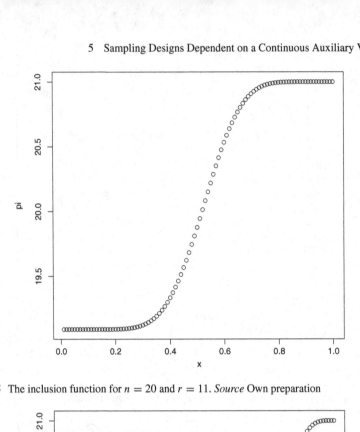

Fig. 5.5 The inclusion function for $n = 20$ and $r = 11$. *Source* Own preparation

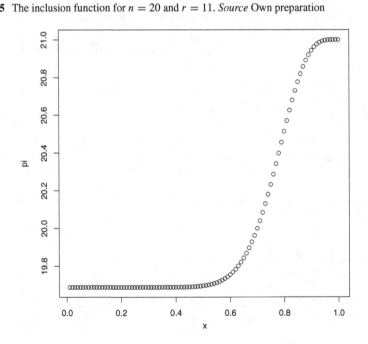

Fig. 5.6 The inclusion function for $n = 20$ and $r = 16$. *Source* Own preparation

Corollary 2 *Theorem 5.7 and expressions (5.43)–(5.45), (5.49) let us derive that when r = 1 and n = 3:*

$$
\begin{cases}
f(x, y, z) = 12x, & \text{for } 0 < x < y < 1 \text{ and } 0 < x < z < 1, \\
f_1(x) = 12x(1 - x)^2, & f_2(x) = f_3(x) = 2x^2(3 - 2x), \\
f_{1,2}(x, y) = f_{1,3}(x, y) = 12x(1 - x), & f_{2,3}(x, y) = 12x^2, \quad x < y, \\
\pi(x) = 4x(3 - 3x + x^2), & \pi(x, y) = 24x(2 - x), \quad x < y.
\end{cases}
$$

If r = 2 and n = 3,

$$
\begin{cases}
f(x, y, z) = 12y, & \text{for } 0 < x < y < z < 1, \\
f_1(x) = 2(1 - 3x^2 + 2x^3), & f_2(x) = 12x^2(1 - x), \quad f_3(x) = 4x^3, \\
f_{1,2}(x, y) = 12g(1 - y), & f_{2,3}(x, y) = 12x^2, \quad f_{1,3}(x, y) = 6(y^2 - x^2), \\
\pi(x) = 2(1 + 3x^2 - 2x^3), & \pi(x, y) = 12(2y - y^2 + x^2), \quad x < y.
\end{cases}
$$

If r = 3 and n = 3,

$$
\begin{cases}
f(x, y, z) = 4z, & \text{for } 0 < x < z < 1 \text{ and } 0 < y < z < 1, \\
f_1(x) = f_2(x) = \frac{4}{3}(1 - x^3), & f_3(x) = 4x^3, \\
f_{1,2}(x, y) = 4(1 - y^2), & f_{1,3}(x, y) = f_{2,3}(x, y) = 4y^2, \\
\pi(x) = \frac{4}{3}(2 - x^3), & \pi(x, y) = 8(1 + y^2), \quad x < y.
\end{cases}
$$

Example 1 When the sample of size $n = 2$ is drawn according to the sampling design defined by expression (5.42) from the uniform distribution on the interval $[0; 1]$, then statistic T_X, given by expression (5.4), is unbiased estimator for parameter $\theta = E(X) = \frac{1}{2}$. Its variance is evaluated based on Corollay 1 and the *Mathematica* program. Therefore, according to expression (5.5), where $g(x) = x$ and $g(x') = y$, we have
  If $r = 1$,

$$
V(T_X) = \frac{1}{3} \int_0^1 \frac{x\,dx}{2 - x} + \frac{4}{3} \int_0^1 \int_0^1 \frac{x\,dx\,dy}{(2 - x)(2 - y)} - \frac{1}{4}
$$

$$
= \frac{4}{3}(ln(2) - 1)^2 + \frac{1}{3}(ln(4) - 1) - \frac{1}{4} = 0.00430966. \tag{5.50}
$$

If $r = 2$,

$$V(T_X) = \frac{2}{3} \int_0^1 \frac{x^2 dx}{x^2 + 1} + \frac{8}{3} \int_0^1 \int_x^1 \frac{xy^2 dx dy}{(x^2 + 1)(y^2 + 1)} - \frac{1}{4}$$

$$= \frac{1}{2}\left(1 - \frac{\pi}{4}\right) + \frac{2}{3}(\pi + 2Catalan - \pi ln(2) + ln(4)) - \frac{35}{12} = 0.349343$$

$$(5.51)$$

where $Catalan = \sum_{k=0}^{\infty}(-1)^k (2k + 1)^{-2}$.

Evaluation of the variance values under several variants of a quantile sampling design requires numerical integration; however, this is not an issue when utilizing a computer package such as *Mathematica*. The evaluated results could easily be generalized into cases when the sampling design depends on a quantile of the uniform distribution on an interval $[0; a]$ where $a > 1$.

5.5 Accuracy Analysis

Let us consider the two-dimensional uniform distribution spanned on the parallelogram, given by the following density function:

$$h(x, y) = \begin{cases} \frac{1}{2\Delta(b-a)}, & (x, y) \in U (a \leq x \leq b, \leq \alpha x + \beta - \Delta \leq y \leq \alpha x + \beta + \Delta), \\ 0, & (x, y) \notin U, \end{cases}$$

$$(5.52)$$

$a, b, \alpha, \beta \in R, \Delta > 0$. The random variable X has the uniform marginal distribution spanned on the interval $[a; b]$. The density function of the marginal distribution of Y is of the trapezoid type and for $b - a \geq 2\Delta$ it is given by the following expression:

$$h(y) = \begin{cases} \frac{1}{2\Delta(b-a)}\left(\frac{y-\beta+\Delta}{\alpha} - a\right), & y \in I_1 = [\alpha a + \beta - \Delta \leq y \leq \alpha a + \beta + \Delta], \\ \\ \frac{1}{b-a}, & y \in I_2 = [\alpha a + \beta + \Delta \leq y \leq \alpha b + \beta - \Delta], \\ \\ \frac{1}{2\Delta(b-a)}\left(b - \frac{y-\beta-\Delta}{\alpha}\right), & y \in I_3 = [\alpha b + \beta - \Delta \leq y \leq \alpha b + \beta + \Delta], \\ \\ 0, & y \notin I_1 \cup I_2 \cup I_3. \end{cases}$$

$$(5.53)$$

When $b - a < 2\Delta$, then in the above expression, the intervals I_1, I_2, and I_3 should be replaced with

$$\begin{cases} I_1' = [\alpha a + \beta - \Delta; \alpha b + \beta - \Delta], \\ I_2' = [\alpha b + \beta - \Delta; \alpha a + \beta + \Delta], \\ I_3' = [\alpha a + \beta + \Delta; \alpha b + \beta + \Delta], \end{cases}$$

respectively. The density function of the conditional distribution is as follows:

$$h(y|x) = \begin{cases} \frac{1}{2\Delta}, & \alpha x + \beta - \Delta \leq y \leq \alpha x + \beta + \Delta, \\ 0, & (x, y) \in R - [\alpha x + \beta - \Delta; \alpha x + \beta + \Delta]. \end{cases} \tag{5.54}$$

The parameters are as follows:

$$\begin{cases} \mu_x = E(X) = \frac{b+a}{2}, \quad V(X) = \frac{(b-a)^2}{12}, \\[2mm] \mu_y = E(Y) = \alpha\mu_x + \beta, \quad E(Y|x) = \alpha x + \beta, \quad V(Y|x) = \frac{\Delta^2}{3}, \\[2mm] V(Y) = V(E(Y|X)) + E(V(Y|X)) = \alpha^2 V(X) + \frac{\Delta^2}{3} = \alpha^2 \frac{(b-a)^2}{12} + \frac{\Delta^2}{3}, \\[2mm] Cov(X, Y) = E((X - \mu_x)E(Y - \mu_y|X)) = E(X - \mu_x)(E(Y|X) - \mu_y)) = \alpha V(X), \\[2mm] \rho = \alpha\sqrt{\frac{V(X)}{V(Y)}}. \end{cases}$$

It is assumed that $\alpha = b = 1$, $a = \beta = 0$. In this case, $\mu_x = \frac{1}{2}$, $V(X) = \frac{1}{12}$, $m_y = \frac{1}{2}$, $V(Y) = \frac{1}{12} + \frac{\Delta}{3}$, and $\rho = \frac{1}{\sqrt{(1+4\Delta^2)}}$. Thus, it can be evaluated that if $\rho = 0.6$, 0.8, 0.9, 0.95, 0.99, then $\Delta = 0.667$, 0.375, 0.242, 0.164, 0.071, respectively. Hence, the marginal density function of Y, defined by (5.53), reduces to the following form:

$$h(y) = \begin{cases} \frac{\Delta+y}{2\Delta}, & -\Delta \leq y < \Delta, \\ 1, & \Delta \leq y \leq 1 - \Delta, \\ \frac{1+\Delta-y}{2\Delta}, & 1 - \Delta < y \leq 1 + \Delta \\ 0, & y < -\Delta \text{ or } y > 1 + \Delta. \end{cases}$$

This density function is shown in Fig. 5.7.

The density function, given by (5.52), of the uniform two-dimensional variable (X, Y) is regarded as a model of the continuous population. In practice, there is an N-element population considered where the two-dimensional variable (X_i, Y_i), $i = 1, \ldots, N$ is attached to each element. Those variables are independent and identically distributed as variable (X, Y). Our goal was to estimate the mean $\bar{Y} = \frac{1}{N}\sum_{i=1}^{N} Y_i$. The following estimators were considered: the ordinary simple random sample mean denoted by $\bar{Y}$, the ordinary ratio and regression estimators denoted by $\bar{Y}_r$ and $\bar{Y}_{reg}$ from the simple random sample, given by expressions (1.9) and (1.14), respectively. The accuracy of these estimators was compared with the accuracy of the following estimator, evaluated based on the continuous quantile sampling design introduced in Sect. 5.4.2 by expression (5.42):

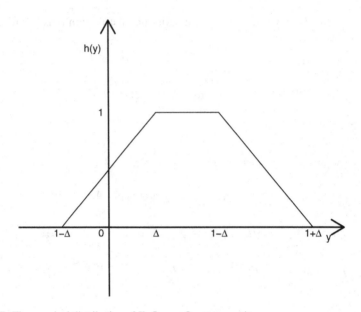

Fig. 5.7 The marginal distribution of Y. *Source* Own preparation

$$T_{Y/X} = \sum_{i=1}^{n} \frac{Y_i}{\pi(x)}$$

where $\pi(x)$ is defined by expression (5.1) (where y should be replaced by x) and Theorem 5.7. This estimator results from the expression (5.11) where, in our case, $h_1(x) = 1$.

The ratio estimator is given by

$$Tr_{Y/X} = \bar{X} \frac{T_{Y/X}}{T_X}, \quad \bar{X} = \frac{1}{N} \sum_{i=1}^{N} X_i, \quad T_X = \sum_{i=1}^{n} \frac{X_i}{\pi(x)}.$$

The regression estimator is as follows:

$$\begin{cases} Tr_{Y/X} = T_{Y/X} + \hat{\alpha}(\bar{X} - T_X), \\[2mm] \hat{\alpha} = \frac{\hat{C}_{XY}}{\hat{V}_X}, \quad \hat{C}_{XY} = \sum_{i=1}^{n} \frac{(X_i - T_X)(Y_i - T_{Y/X})}{\pi(X_i)}, \quad \hat{V}_X = \sum_{i=1}^{n} \frac{(X_i - T_X)^2}{\pi(X_i)}. \end{cases}$$

These estimators were evaluated based on the quantile sample and the inclusion function derived in Sect. 5.4.2.

The simulation procedure for investigating the estimation accuracy is as follows. At the tth ($t = 1, \ldots, M$) step of the algorithm, N independent observations of the two-dimensional random variable (X, Y) are generated according to the uniform density function defined by expression (5.52). In this case, observations of X are randomly and independently generated according to the uniform distribution defined on the interval $[0; 1]$. Next, the value y_i of Y is generated according to the conditional distribution given by expression (5.54), $i = 1, \ldots, N$. Table 5.1 is evaluated based on the data generated for the parameters $a = 0$, $b = 1$, $\alpha = 1$, and $\beta = 0$, while Table 5.2 is evaluated based on the data generated for the parameters: $a = 0$, $b = 1$, $\alpha = -1$, and $\beta = 1$. Figures 5.8 and 5.9 show domains of these two distributions. Next, a sample of size n is drawn from the sequence $\{x_1, \ldots, x_N\}$ according to the quantile sampling scheme explained by expressions (5.39)–(5.41). Finally, the values of the estimators, which were considered, are calculated. When the algorithm is terminated, the following relative efficiency coefficient is evaluated:

$$deff(T) = \frac{\hat{V}(T)}{\hat{V}(\bar{Y})}, \quad \hat{V}(T) = \frac{1}{M}\sum_{t=1}^{M}(T_t - \bar{T})^2, \quad \bar{T} = \frac{1}{M}\sum_{t=1}^{M}T_t.$$

Table 5.1 The relative efficiency coefficients of the strategies. The uniform distribution $U((x, y) : a \leq x \leq b, x - \Delta \leq y \leq x + \Delta)$, $N = 3000$. The number of iterations: 10,000

n	r	$T_{Y/X}$			$T_{rY/X}$			$T_{regY/X}$			$\bar{Y}_r$			$\bar{Y}_{reg}$		
		0.80	0.90	0.95	0.80	0.90	0.95	0.80	0.90	0.95	0.80	0.90	0.95	0.80	0.90	0.95
1	2	3	4	5	6	7	8	9	10	11	12	13	14	16	16	17
3	1	54	39	32	45	24	13	44	24	13						
	2	45	32	27	29	21	16	35	39	18	65	34	17	345	149	63
	3	54	46	38	26	15	7	67	39	17						
5	2	53	45	56	19	14	10	23	21	14						
	3	42	43	55	14	13	10	23	23	17	47	25	13	61	31	16
	4	57	65	60	13	19	11	44	43	22						
10	3	64	59	54	39	40	7	65	54	14						
	6	64	55	53	25	13	15	56	29	24	41	22	11	42	22	12
	8	66	62	53	22	17	7	60	38	21						
20	6	72	64	87	22	13	9	53	29	21						
	11	74	66	85	24	13	33	58	34.2	46	39	21	10	40	21	11
	16	76	67	88	22	13	11	60	33	25						

Source Own calculations

Table 5.2 The relative efficiency coefficients of the strategies. The uniform distribution $U((x, y) : a \leq x \leq b, 1 - x - \Delta \leq y \leq 1 - x + \Delta)$, $N = 3000$. The number of iterations: 10000

		$T_{Y/X}$			$Treg_{Y/X}$			$\bar{Y}_{reg}$		
n	r	0.80	0.90	0.95	0.80	0.90	0.95	0.80	0.90	0.95
1	2	3	4	5	6	7	8	9	10	11
	1	363	684	770	408	819	999			
3	2	106	116	123	141	129	283	345	149	63
	3	94	86	90	116	89	482			
	2	99	110	107	77	104	77			
5	3	94	97	98	80	127	164	57	31	16
	4	82	80	85	55	58	80			
	3	89	91	84	63	85	34			
10	6	85	79	83	86	50	18	38	22	12
	8	82	79	80	64	71	60			
	6	81	83	80	81	33	125			
20	11	83	80	80	61	50	139	39	21	11
	16	86	77	77	21	58	99			

Source Own calculations

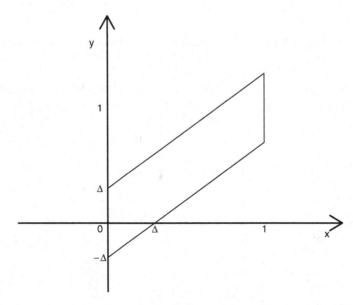

Fig. 5.8 The support of the uniform distribution for $a = 0$, $b = 1$, $\alpha = 1$, and $\beta = 0$. *Source* Own preparation

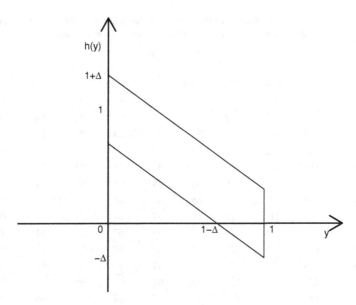

Fig. 5.9 The support of the uniform distribution for $a = 0$, $b = 1$, $\alpha = -1$, and $\beta = 1$. *Source* Own preparation

Variance $\hat{V}(\bar{Y})$ is evaluated separately based on the independently replicated simple random samples of size n drawn without replacement from $(Y_1, \ldots, Y_N)$. The bias of an estimator T is defined as the following ratio:

$$b(T) = \frac{\bar{T}}{\bar{Y}} 100\%, \quad \bar{\bar{Y}} = \frac{1}{M} \sum_{t=1}^{M} \bar{Y}_t, \quad \bar{Y}_t = \frac{1}{N} \sum_{i=1}^{N} Y_i.$$

Statistic T is the unbiased estimator of $\bar{Y}$, if $b(T) = 100\%$. When $b(T) > 100\%$ ($b(T) < 100\%$), then T overestimates (lower-estimates) the mean $\bar{Y}$.

The relationship between the rank r of the order statistic $X_{(r)}$ and the rank p of the sample quantile is as follows:

$$Q_p = X_{(r)}, \quad r = [np] + 1$$

where the function $[np]$ is equal to the largest integer not greater then np.

Table 5.1 is prepared under the assumed support of the variables (X, Y) shown in Fig. 5.8. An analysis of this table leads to the following conclusions. In the case of all considered types of the quantile sampling design, the ratio $Tr_{Y/X}$ and regression $Treg_{Y/X}$ estimators are superior to $T_{Y/X}$. The ratio $Tr_{Y/X}$ estimator from the quantile sample is not less accurate than $\bar{Y}_r$ the ratio estimator from the simple random sample drawn without a replacement, however, this is not the case when the sample size $n = 20$. Estimator $Treg_{Y/X}$ is superior to $\bar{Y}_{reg}$ for $n \leq 5$. The precision of all the estimators is improved when the correlation coefficient between the variable under study and the auxiliary variable increases. In general, when the sample size increases,

Table 5.3 The relative bias coefficients of the strategies. The uniform distribution $U((x, y) : a \leq x \leq b, \ \alpha x + \beta - \Delta \leq y \leq \alpha x + \beta + \Delta)$, $N = 3000$. The number of iterations: 10000

| | | $\alpha = 1, \ \beta = 0$ | | | $\alpha = -1, \ \beta = 1$ | | | | | |
| | | $T_{Y/X}$ | | | $T_{Y/X}$ | | | $Treg_{Y/X}$ | | |
n	r	0.80	0.90	0.95	0.80	0.90	0.95	0.80	0.90	0.95
1	2	3	4	5	6	7	8	9	10	11
	1	107	107	107	74	75	75	75	76	76
3	2	113	113	113	70	70	69	75	75	74
	3	122	122	122	70	69	69	80	82	83
	2	115	116	116	67	67	68	77	76	77
5	3	119	119	119	68	68	68	81	81	81
	4	120	122	123	68	69	68	87	86	87
	3	116	119	123	66	67	68	87	86	86
10	11	125	116	126	67	67	67	91	91	92
	8	128	128	128	68	67	68	95	95	95
	6	128		128	66	67	66	94	93	93
20	11	129	129	129	67	67	67	96	96	96
	16	131	131	131	67	67	67	98	98	98

Source Own calculations

the relative efficiency coefficient for the estimators based on the quantile sample also increases. Therefore, the estimators from the quantile samples are especially useful for small sample sizes. In general, estimator $T_{Y/X}$ is the most accurate for the rank r which identifies the sample median of the random variable X. This tendency is similar for the estimator $Tr_{Y/X}$.

Besides the statistic $T_{Y/X}$, all the estimators considered in Table 5.1 are not substantially biased because $99.5\% \leq b(T) \leq 100.4\%$. Therefore, in Table 5.3 there are presented only biases of $T_{Y/X}$. This statistic overestimates $\bar{Y}$. The levels of the relative bias are similar for all considered values of the correlation coefficient $\rho = 0.8; 0.9; 0.95$. There is also a slight tendency for the bias to increase as the sample size increases.

Table 5.2 is prepared under the assumed domain of the variable (X, Y) distribution shown in Fig. 5.9. Estimator $Tr_{Y/X}$ was omitted because the ratio estimator is not suitable for the situation when the intercept coefficient of the regression function does not equal zero. In this case, variables X and Y are negatively correlated. An analysis of this table leads to the following conclusions. In general, when $n = 3$ the accuracies of all the estimators from the quantile sample are not superior to the simple random sample mean. Moreover, the regression estimator from the simple random sample is superior to $T_{Y/X}$ and $Treg_{Y/X}$. An analysis of Table 5.3 shows that statistics $T_{Y/X}$ and $Treg_{Y/X}$ underestimate the mean $\bar{Y}$. The biases of the regression estimator $\bar{Y}_{reg}$ from the simple random sample drawn without a replacement were omitted in Table 5.3 because they were negligible ($98.8\% \leq b(\bar{Y}_{reg}) \leq 100.5\%$). Moreover, levels of $b(T_{Y/X})$ bias were similar for all the sample sizes considered as well as the values of the correlation coefficient, and they oscillated from 66 to 74%. There was a slight tendency for the bias of $Treg_{Y/X}$ to decrease when the sample size increases.

In conclusion, when the uniform two-dimensional distribution, given by expression (5.52), has the intercept coefficient of the linear regression equal to zero, the estimator from the quantile sample could compete with the ratio and regression estimators from the simple random sample, but in the case of a small sample size.

The results of this analysis demonstrate that future research on new continuous sampling designs could provide valuable estimation procedures from a theoretical and practical point of view.

References

Bąk, T. (2014). Triangular method of spatial sampling. *Statistics in Transition, 15*(1), 9–22. http://stat.gov.pl/en/sit-en/issues-and-articles-sit/

Bąk, T. (2018). An extension of Horvitz-Thompson estimator used in adaptive cluster sampling to continuous universe. *Communications in Statistics-Theory and Methods, 46*(19), 9777–9786. https://doi.org/10.1080/03610926.2016.1218028

Benhenni, K., & Cambanis, S. (1992). Sampling designs for estimating integrals of stochastic processes. *The Annals of Statistics, 20*(1), 161–194.

Bucklew, J. A. (2004). *Introduction to rare event simulation*. New York, Berlin, Heidelberg, Hong Kong, London, Milan, Paris, Tokyo: Springer.

Cordy, C. B. (1993). An extension of the Horvitz-Thompson theorem to point sampling from a continuous universe. *Statistics and Probability Letters, 18,* 353–362.

Cox, D. R., & Snell, E. J. (1979). On sampling and the estimation of rare errors. *Biometrika, 66*(1), 125–32.

Cressie, N. A. C. (1993). *Statistics for spatial data*. New York: Wiley.

David, H.A., & Nagaraja, H.N. (2003). *Order statistics*. Wiley.

Frost, P. A., & Tamura, H. (1986). Accuracy of auxiliary information interval estimation in statistical auditing. *Journal of Accounting Research, 24,* 57–75.

Ghirtis, G. C. (1967). Some problems of statistical inference relating to double-gamma distribution. *Trabajos de Estatistica, 18,* 67–87.

Horvitz, D. G., & Thompson, D. J. (1952). A generalization of the sampling without replacement from finite universe.

Kotz, S., Balakrishnan, & Johnson, N. L. (2000). *Continuous multivariate distributions, vol. 1: models and applications*. Wiley, New York, Chichester, Wenheim, Brisbane, Sigapore, Toronto.

McKay, A. T. (1934). Sampling from batches. *Journal of the Royal Statistical Society, 2,* 207–216.

Ripley, B. D. (1987). *Stochastic simulation*. Wiley, New York.

Thompson, M. E. (1997). *Theory of sample survey*. London, Weinheim, New York, Tokyo, Melbourne, Madras: Chapman & Hall.

Wilhelm, M., Tillé, Y., & Qualité, M. (2017). Quasi-systematic sampling from a continuous population. *Computational Statistics & Data Analysis, 105,* 11–23.

Wywiał, J. L. (2016). *Contributions to testing statistical hypotheses in auditing*. Warsaw: PWN.

Wywiał, J. L. (2018). Application of two gamma distribution mixture to financial auditing. *Sankhya B, 80*(1), 1–18.

Wywiał, J. L. (2020). Estimating the population mean using a continuous sampling design dependent on an auxiliary variable. *Statistics in Transition new series, 21*(5), 1–16. https://doi.org/10.21307/stattrans-2020-052; https://sit.stat.gov.pl/Article/157

Zubrzycki, S. (1958). Remarks on random, stratified and systematic sampling in a plane. *Colloquium Mathematicum, 6,* 251–262. https://doi.org/10.4064/cm-6-1-251-264; http://matwbn.icm.edu.pl/ksiazki/cm/cm6/cm6135.pdf

Printed in the United States
by Baker & Taylor Publisher Services